P...
Tec...

Life Cycle Assessment

Life Cycle Assessment addresses the dynamic and dialectic of building and ecology, presenting the key theories and techniques surrounding the use of life cycle assessment data and methods.

Architects and construction professionals must assume greater responsibility in helping building owners to understand the implications of making material, manufacturing, and assemblage decisions and therefore design to accommodate more ecological building. *Life Cycle Assessment* is a guide for architects, engineers, and builders, presenting the principles and art of performing life cycle impact assessments of materials and whole buildings, including the need to define meaningful goals and objectives and critically evaluate analysis assumptions.

As part of the *PocketArchitecture* series, the book includes both fundamentals and advanced topics. The book is primarily focused on arming the design and construction professional with the tools necessary to make design decisions regarding life cycle, reuse, and sustainability. As such, the book is a practical text on the concepts and applications of life cycle techniques and environmental impact evaluation in architecture and is presented in language and depth appropriate for building industry professionals.

Kathrina Simonen is an Assistant Professor in the Department of Architecture at the University of Washington. Licensed as both an Architect and Structural Engineer, her research agenda stems from unresolved questions originated during over 15 years of professional practice. Her research topics include life cycle assessment (LCA) and practice innovations such as prefabrication, digital manufacturing, and alternative project delivery models. She is founding director of the Carbon Leadership Forum, an industry/academic research collaborative focused on linking the science of life cycle assessment to industry best practices in order to help enable quantifiable reduction of the environmental impact of the built environment.

PocketArchitecture:
Technical Design Series

Series Editor: Ryan E. Smith

Building Information Modeling
Karen M. Kensek

Life Cycle Assessment
Kathrina Simonen

Daylighting and Integrated Lighting Design
Christopher Meek and Kevin Van Den Wymelenberg

Architectural Acoustics
Ana M. Jaramillo and Chris Steel

Contents

Preface

As a practising architect, I struggled to understand the environmental impact of building materials and strove to find ways to integrate this knowledge into design and construction practice. What began as a search for the "carbon footprint" of specific materials and transportation methods expanded to a study of the broad range of environmental impacts related to building materials, products and buildings using a life cycle perspective.

As an architecture researcher, I have been exploring life cycle assessment (LCA) methods, data and tools to understand the opportunities and challenges that these resources provide to architects (and others in the building industry) looking to understand and reduce the environmental impact of buildings.

This book is designed to be the resource that I searched for when I began attempting to use LCA data in practice: an introduction to the fundamentals of LCA, graphic representation of abstract concepts, developed examples of application to the building industry, and identification of relevant LCA resources.

Acknowledgments

My understanding of LCA has been greatly enhanced by the support from fellow faculty members including, most significantly, Joyce Cooper, Associate Professor of Mechanical Engineering at the University of Washington, who graciously included me (and my student researchers) in her LCA graduate seminar and has provided sound advice and mentorship. Additionally, Liv Haselbach, Associate Professor of Civil and Environmental Engineering at Washington State University, and Elaine Oneil, Research Scientist in the College of the Environment at the University of Washington, along with Joyce Cooper, collaborated effectively to develop research exploring the integration of LCA into the Washington State Building code. The research we developed together for that report provided a solid foundation from which to develop the content of this book. Thank you, Joyce, Liv and Elaine.

Thank you also to graduate student researchers: Kyle Boyd, Yasaman Esmaili, David Fish, Dahra Goradia, Lissa Gotz, Monica Huang, Jocelyn Reutebuch, Kristen Strobel and Mazohra Thami. Their LCA research contributed both directly and indirectly to this book. Mazohra deserves special recognition for her help in establishing the graphic quality of the illustrations.

Generous permission to use or adapt images, text and/or data was provided by the following: Armstrong World Industries, Arup, the Carnegie Mellon University Green Design Institute (CMU GDI), CEDA, the Consortium for Research on Renewable Industrial Materials (CORRIM), the European Commission Joint Research Centre Institute for Environment and Sustainability, the European Environment Agency (EEA), the European Committee for Standardization (CEN), the International Organization for Standardization (ISO), John Basbagil, the *Journal of Environmental Science and Technology*, the *Journal of Industrial Ecology*, the Forestry Commission (UK), the National Oceanic and Atmospheric Administration (NOAA), the Institute of Construction and Environment (IBU), National Renewable Energy Laboratory (NREL), UL Environment and the World Resource Institute (WRI).

I would also like to thank the following LCA practitioners and industry experts who took the time to read sections of this book including: Claire Broadbent, Lindita Bushi, Chris Erickson, Erin Moore, Elaine Oneil, Tien Peng, Edie Sonne Hall, Cassie Theil and Frances Yang. Their comments greatly improved the clarity and accuracy of the text and images.

I am pleased to be included in this new series of handbooks for architects and appreciate the editorial advice of Ryan Smith and the enthusiasm of the staff at Routledge.

Support for the research and development for the book was provided by the Mithun/Russell Family Foundation Endowed Professorship in Sustainability, the University of Washington's College of Built Environments and the Carbon Leadership Forum (www.carbonleadershipforum.org) industry sponsors who include: Arup, Central Concrete, the National Ready Mixed Concrete Association, Webcor Builder as well as Climate Earth, Ceratech, Degenkolb and Magnusson Klemencic Associates.

And most significantly, I would like to thank my mother, Candace, for reading, editing and improving my writing from the beginning; my father, Tom, for encouraging me to pursue engineering; Oliver, Max and Gus for nighttime NERF battles; and Rolland for his love and support.

Abbreviations

ANSI	American National Standards Institute
ASTM	ASTM Standards
BOF	Basic oxygen furnace
CEN	European Committee for Standardization
EAF	Electric arc furnace
EEA	European Environmental Agency
EIO	Economic input–output
EPA	Environmental Protection Agency (US)
EPD	Environmental Product Declaration
EU	European Union
GCV	Gross calorific value
GGBFS	Ground granulated blast furnace slag
GHG	Greenhouse gas
GLB	Glue laminated beam (or timber)
HPD	Health product declaration
ISO	International Standards Organization
LCA	Life cycle assessment
LCCA	Life cycle cost analysis
LCI	Life cycle inventory
LCIA	Life cycle impact assessment
MEP	Mechanical Electrical and Plumbing
NCV	Net calorific value
NO_x	Nitrogen oxides
PAS	Publicly Available Specification from British Standards Institution
PCR	Product category rules
PM	Particulate matter

SETAC	Society for Environmental Toxicology and Chemistry
SO_x	Sulfur oxides
TRACI	Tool for the Reduction and Assessment of Chemical and Other Environmental Impacts (US EPA)
UN	United Nations
UNEP	United Nations Environment Program
US	United States

chapter 1

Introduction

ALL BUILDING RESULTS IN ENVIRONMENTAL IMPACTS. Emissions related to energy generation and manufacturing pollute the air we breathe, impact global climate and impact the health of animals and plants. Local environments are changed when land is re-shaped and vegetation is replaced with construction. The challenge of developing truly sustainable or even regenerative buildings (Cole, 2012) has led to a desire to understand building and construction from a systems-based perspective. Buildings are not static objects, but rather one component within complex environmental, social and economic systems.

Currently, "green" building practices strive to do less harm than conventional building methods. In many instances, single attributes such as percentage of recycled content or locally sourced materials, are used to identify a product as environmentally preferable. However, sophisticated users can imagine that these single attributes may not capture the total environmental picture, for example, a local product sourced in an inefficient and polluting facility might have larger environmental impact than one produced farther away in an efficient factory and shipped to a site.

Understanding a building or product from the perspective of its entire life cycle is the first step in developing sustainable and regenerative systems. How can the net impact be positive? How does one component fit into a larger system? Arguably, understanding the impacts throughout the life cycle is essential for assessing if a product or building is "green." If we cannot determine what the impacts are, how can we be sure that we have reduced them?

Life Cycle Assessment (LCA) is a standardized method of tracking and reporting the environmental impacts of a product or process throughout its full life cycle (ISO, 2006a: 8). Originally developed from principles of industrial ecology (Jelinski et al., 1992; Guinee et al., 2011) and first applied to the manufacturing of products within a factory (Hunt and Franklin, 1996), LCA methods and data are increasingly being used to evaluate the materials and products used in building construction and to assess a whole building from construction to end of life. Understanding the strengths and weaknesses of different LCA methods and different LCA assumptions is critical for proper interpretation and use of LCA.

A simplified diagram of the life cycle stages of a building is shown in Figure 1.1. In order to assess the total environmental impact of a building, all life cycle phases must be considered, from material extraction, manufacturing, construction, use (operations, maintenance and refurbishment) through eventual demolition and disposal. Note that some LCA data is reported as cradle-to-gate and thus only includes impacts related to manufacturing of a product up to the "gate" of the factory. A comprehensive LCA considers impacts from cradle-to-grave. A cradle-to-cradle analysis would track how products at end of life become material resources for other products (McDonough *et al.*, 2002).

1.1 Simplified life cycle stages of a building

LIFE CYCLE ASSESSMENT IS NOT the same as life cycle costing, nor does it capture all environmental impacts well. Life cycle cost analysis (LCCA) tracks the financial implications of different options including first

1.2 Life cycle tracking of emissions to and extractions from nature

costs, operating costs and refurbishment/replacement costs. LCCA is not covered in this reference. LCA tracks the quantities of emissions to nature (e.g. kg of carbon dioxide and methane) and extractions from nature (e.g. kg of iron ore) for a studied product or process throughout its life cycle as represented in Figure 1.2.

LCA reports impacts that can be simply and predictably measured. Thus impacts from fuel combustion and process chemical emissions are represented more effectively than local impacts such as surface water runoff or habitat disruption (see Chapter 3).

A life cycle approach to buildings requires an understanding that buildings are not static objects "finished" when construction stops and an owner occupies. Rather, buildings are dynamic changing entities that require an understanding of their impact on society and the environment throughout their life. Building operations account for a staggering proportion of the global annual energy consumption. Accordingly, increasing the energy efficiency of new and retrofit buildings has been a strong focus of the building industry. Building construction and renovation account for a sizeable minority of these energy and environmental impacts. LCA provides methods to quantify these impacts. While detailed information about methods to quantify operational energy use can be found elsewhere, LCA treats operational impacts as one life cycle stage to be evaluated.

Depending on the particular building in question, the environmental impacts related to building materials, construction, maintenance and end of life (termed embodied impacts) can range between nearly 0 per cent to nearly 100 per cent. For example, a passive house with site-generated renewable operational energy would have zero operational impacts and thus the embodied impacts would be 100 per cent. The embodied impacts tent in the attic heated by coal would be near zero. There is not yet conclusive data to determine the typical ratio of impacts between construction and operation (Moore, 2013). An LCA study comparing variations on a typical theoretical building (Basbagill, 2013) demonstrate that for typical buildings, these embodied impacts account for between 10–20 per cent of the total impacts, and operational impacts account for 80–90 per cent. A recent French study

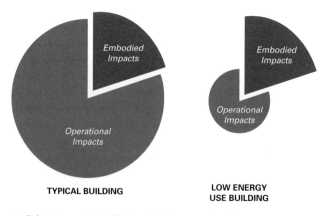

TYPICAL BUILDING   **LOW ENERGY USE BUILDING**

1.3 Relative impact of a building's embodied vs. operational impacts that compared 70 different actual case study buildings found the embodied impacts accounted for an average of 19 per cent when assuming a 100-year life span (HQE, 2012). However, as shown in Figure 1.3, the embodied impacts become increasingly more significant as the operational efficiency improves.

LCA can provide the analytical framework to identify environmental impacts, improve manufacturing processes and compare between alternatives. LCA provides quantifiable metrics by which to assess the environmental impacts of a product or process, helping to avoid generalized statements and ideally can help avoid "green-washing".

1.1 LCA: environmental accounting

LCA IS A METHOD OF ENVIRONMENTAL ACCOUNTING: tracking the inputs from nature (such as limestone, water and coal) and outputs to nature (such as waste, carbon dioxide and methane) considering all of the processes that take place during the manufacture, use and disposal of a product (or system). Figure 1.4 represents how each stage of a building requires energy and material inputs and outputs wastes and emissions. LCA tracks these inputs and outputs. As an accountant can use the "cash" or "accrual" method to track a

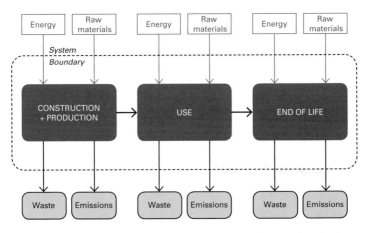

1.4 LCA tracks inputs from nature and outputs to nature that cross the "system boundary"

company's economic performance and attain different results, there are different methods that can be used when performing an LCA.

A life cycle inventory (LCI) is a detailed accounting of the quantities of raw materials used, the products produced, the waste outputs and the emissions to air, land and water. See Chapters 2 and 4 for more details on the generation of LCI data. In more complex processes, hundreds or thousands of emissions to nature are tracked. In order to simplify interpretation of this data, the emissions are amalgamated into summary environmental impacts categories (such as climate change or acidification).

Ideally, an LCA would report on "all" relevant environmental impacts. Practically, however, LCA reports impacts that are straightforward to calculate, given the existing LCI data, are calculated with methods that are broadly supported, and are globally relevant. In current LCA practice, this is limited to between five and eight primary environmental impacts along with consumption/emission quantities for items such as water, energy and waste. Chapter 3 provides definitions and information about these categories and methods for calculating these impacts are listed in Figure 1.5. See Chapter 3 for additional environmental impacts tracked by LCA.

Evaluating the life cycle impacts reported by an LCA requires knowledge of LCA principles to interpret and apply the results to improve processes and/or make choices. Figure 1.6 shows the LCA results comparing four different wall assembly options for a hypothetical office building. Note, wall type 4 performs best in the three categories of fossil fuel consumption, carbon footprint and acidification but significantly worse in ozone depletion and smog creation. One might decide to prioritize one impact over another or alternately might select option 2, which is consistently "good" even if not the "best" in some categories.

IMPACTS	INVENTORY
Established Methodology	• Water Consumption
• Acidification	• Energy Use
• Climate Change/Global Warming	• Waste Generation
• Eutrophication	
• Ozone Depletion	
• Photochemical Ozone Creation/Smog	

1.5 Typical impacts tracked by LCA

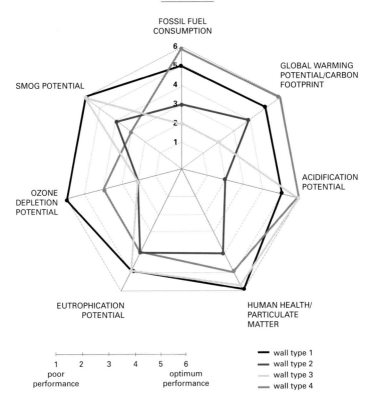

1.6 Hypothetical life cycle impact results comparing four assembly types

Some policy-makers wish to have a single environmental performance "score" to enable them to rank choices. In order to do this, LCA results can be weighted by relative importance. The US National Institute of Standards and Technology provides a rating method in its LCA database and tool, BEES (NIST, 2013), and permits users to over-ride this with their own priorities. There are other methods available that normalize and weight LCA impacts. There is no global consensus on an appropriate method to combine impacts; further, current practice tends to report all LCA results separately.

1.2 LCA: strengths and weaknesses

ALTHOUGH LCA DATA AND METHODS PROVIDE QUANTIFIABLE measures of environmental impact, cautious and critical evaluation of the capabilities and culpabilities of LCA is required when interpreting and acting on the results of LCA. Users of LCA data should become familiar with the methodological challenges (see Chapter 6) in order to ensure that they are able to understand the background of LCA data and assess the relevance of inclusion within their own analysis.

What follows is a short description of the key strengths and weaknesses of LCA.

LCA strengths

- ■ Quantifiable: LCA provides quantifiable metrics by which to evaluate environmental impacts. LCA standards provide guidance on how to develop a comprehensive and systematic evaluation.
- ■ Comparable: If the analysis is framed *correctly*, LCA results can be used to compare between different options. Understanding the requirements for "comparative assertions" is key for applying LCA appropriately. See Section 5.1 for more information.
- ■ Comprehensive: An LCA can track impacts throughout the supply chain and throughout the life cycle giving a more complete understanding of environmental impacts than can be attained by other more limited evaluation methods.
- ■ Indicative: An LCA can provide insights to manufacturers and consumers looking to understand opportunities for improvement and identify where trade-offs occur from one life cycle phase to another or from one impact to another.
- ■ Motivational: An LCA provides insights to manufacturers and consumers that can motivate improvements.
- ■ Avoids green-washing: The quantitative results of LCA can help to minimize the use of unsubstantiated environmental claims such as "greener", "low impact", etc.

LCA weaknesses

■ Time-consuming: A detailed LCA can require significant resources to execute: knowledge of the product and processes, time to assemble the data, access to LCI databases and expertise in evaluating and computing LCA results. Little research has been done on establishing the cost-to-benefit ratio of performing LCAs.

■ Incomplete: LCA reports environmental impacts that are able to be quantified using current methods. Global and regional impacts related to emissions from fuel combustion are easier to track than local impacts such as habitat disruption or indoor air quality. Although LCA standards advise tracking qualitative environmental impacts, in practice, this is significantly more difficult to implement. Additionally, the social and economic impacts of sustainability are not yet adequately considered in standardized LCA methods to be effective.

■ Incomplete data: Data sets often do not exist for the specific location and process being studied. Therefore, proxy data sets from different regions are often used to replace missing data. Much LCA data is proprietary and thus not available to those who have not purchased access to the data via LCA software.

■ Requires judgement: There are many aspects of a LCA study that require the practitioner preparing the analysis to exercise judgement: what to include, which data to use, how to model recycling, etc. These methodological challenges are outlined in more detail in Chapter 6. Additionally, LCA results can be inconclusive. For example, one option may have lower acidification while another has lower water use. Determining how to interpret and act on this data requires knowledge and judgement.

■ Uncertain: A full cradle-to-grave LCA typically requires the development of scenarios to outline the use and end of life phases. For example, changing the assumed life span of a building can significantly change which life cycle stages are the dominant environmental impacts. Actual events such as natural disasters, renovations and material re-use strategies are difficult to model, and changes in assumptions can also significantly change LCA results.

■ Faulty precision: As LCA results often report the aggregation of industry average data without in-depth consideration of the variability and uncertainty, yet still report data to four decimal points, users can be misled to believe that the results are more precise than they actually are. LCA results are best used to compare within consistently structured analyses and less relevant when comparing across studies with different assumptions, different data sources and different methods.

■ **Disguises green-washing:** If LCA results are presented based on undocumented and/or unusual methods (especially without the inclusion of other known environmental benefits and without the context to help the user interpret data), the data of LCA could be used to inaccurately or incompletely support claims of environmental preference or transparency.

1.3 LCA in the building industry

THERE IS GROWING INTEREST IN INTEGRATING the methods and data of LCA into the evaluation of the environmental impacts of building materials, building products, as well as those of whole buildings. Europe is more advanced in developing LCA methods, data and applications than elsewhere. LCA is more commonly used as a research tool to analyze buildings or manufacturing processes than to aid design decision-making. At present, there are four main applications of LCA in the building industry: (1) evaluating manufacturing processes; (2) developing product labels; (3) comparing materials and methods; and (4) analyzing whole buildings.

CASE STUDY 1.1: USES OF LCA IN THE BUILDING INDUSTRY
The following are examples of ways in which LCA can be used in the building industry:

■ Evaluating manufacturing: A manufacturer, a group of manufacturers or research organization can commission an analysis of their manufacturing process. These analyses may only focus on gate-to-gate impacts that a specific manufacturer can control. Often these studies report the inventory of emissions for the process under study, developing a LCI that can be integrated into a larger LCA. For example, the emissions related to producing reinforcing steel are reported in the US LCI database. This information could be integrated into a whole building LCA of an office building. See Sections 2.3 and Chapter 4 for more information about inventories.

■ Developing product declarations: Environmental Product Declarations (EPDs) are standardized reports of LCA results. Idealized as an environmental "nutrition" label for products, EPDs are typically multi-page reports that include information on the LCA data, methods and assumptions as well as reporting the LCA results for a specific product following standardized rules (Product Category Rules or PCRs) for computing the LCA. EPDs are discussed in more detail in Section 5.3.

■ Comparing materials and methods: Many LCA studies have been performed to compare a material (e.g. a steel or concrete structure) or a method (precast vs. cast in place concrete). Often funded by industry trade organizations, these studies look to identify an environmentally preferable method. Based on established LCA standards, LCA data should only be used to make decisions between options (termed comparative assertions) unless very specific analysis requirements are met (see Section 5.2).

■ Analyzing buildings: A whole building LCA can assess the impacts of manufacturing and construction in context with the operations and maintenance impacts. These whole building LCAs can be performed to evaluate methods as discussed above, to develop an understanding of the largest drivers of environmental impacts and identify opportunities for improvement. Government agencies and green building rating systems are beginning to test the potential of using LCA to evaluate whole building performance. In order to analyze a building, assumptions must be made about its life. The establishment of building life, including its use, maintenance, refurbishment and end-of-life treatment is all uncertain and can be made based on conjecture, statistical data for past buildings or testing variable alternatives. Modeling assumptions should match current practice or evaluate the impacts of different scenarios. Sections 6.5 and 6.6 go into these aspects in more detail.

LCA DATA IS JUST ONE MEASURE OF PERFORMANCE that should be added to the multiple quantitative and qualitative performance criteria by which we evaluate buildings, and must be placed in context within a larger decision-making framework. Figure 1.7 demonstrates the multiple criteria that could be assessed when comparing different options for developing an office building. In this case, four different design options are studied. The standard office

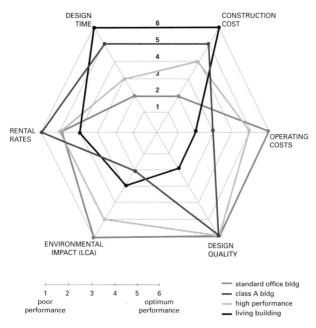

1.7 Integrating LCA data into design decision-making

building optimizes first costs and performs worst in operating costs and design quality. A high performance building may require larger initial construction costs but perform well in reducing operating costs. Design and construction decisions require constant assessment of (sometimes conflicting) criteria. Trade-offs are often made, optimizing one performance attribute while recognizing that other attributes are compromised. Note, from this perspective "LCA impacts" are evaluated as a single performance criterion. As demonstrated in Figure 1.6, evaluating an LCA already requires balancing multiple performance criteria.

As noted earlier, developing a comprehensive LCA can require a significant investment to collect and analyze data. Buildings are complex products and the "scope", scale and detail of analysis, (see Section 2.2) can vary, depending upon the "goal" or objectives (see Section 2.1). LCA can analyze buildings, building components, building materials and building processes. Figure 1.8 represents four different LCA studies that could be performed for

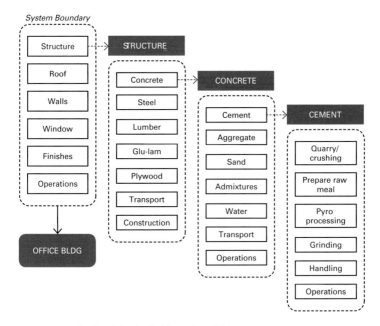

1.8 Variations of scale and detail in building industry LCAs

a whole building, parts of the building, and/or building products. At each of these scales a different level of data refinement is required and care must be taken to match an appropriate depth of analysis to the intent.

LCA IS AN EMERGING DISCIPLINE and the integration of LCA into building design, construction and manufacturing practices is likely to develop quickly in the near future. This book is designed as a reference source for architects, engineers and builders, who are looking to better understand how LCA data and methods can be used to assess and inform the buildings they design and construct. The content is focused on LCA fundamentals to offer building industry professionals the background to be able to evaluate LCA data, explore the integration of LCA methods and help lead the industry towards the needed reductions of the environmental impacts of buildings. Chapter 7 identifies additional resources for data and tools and references to provide the additional technical guidance needed for those wishing to perform their own LCA studies.

chapter 2

Life Cycle Assessment Fundamentals

LIFE CYCLE ASSESSEMENT METHODOLOGY evolved from efforts to evaluate and improve the manufacturing processes of products during the 1960s and 1970s, initially focusing on the tracking of energy and resource use. The methods were formalized through the development of LCA standards by organizations such as the International Standards Organization (ISO) in the 1990s. *The LCA fundamentals outlined in this book are an interpretation of the requirements that are explicitly defined by regional and international standards. Readers looking to develop their own LCAs should obtain copies of current LCA standards as these will include specific language and definitions as well as expanded detail for the requirements that are not captured in this overview.* See Chapter 5 for more detailed information on LCA standards.

Note, the language of LCA has specific meanings defined by LCA standards. While the general definitions of LCA terms are expanded and explained within the text (and defined in the Index of Standard Definitions on p. 149), the precise word definitions are copyrighted by the standards bodies and not included in this book. The terms that are defined in the standards are noted in the Index of Standard Definitions and the appropriate standards are referenced.

The basic phases in an LCA include: goal and scope definition; inventory analysis; impact assessment; and interpretation. The double-headed arrows in Figure 2.1 indicate the iterative approach within and between the phases that is required to ensure that goals, methods and conclusions are internally consistent, comprehensive and relevant. The results (potential environmental impacts) of a LCA study will vary depending upon the goal (why a study is being performed), and the scope (what is included in the study). Thus careful definition of the goal and scope is a critical first step in performing an LCA. Establishing which LCA stages (as discussed in Chapter 1 and Section 2.2) are included in an LCA study is a critical component of

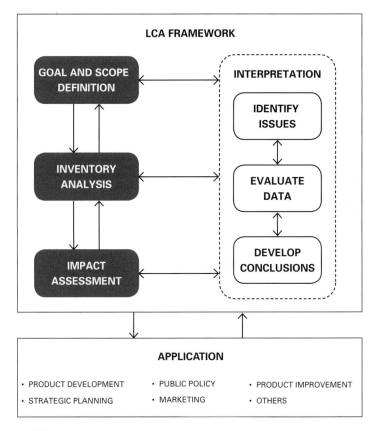

2.1 LCA phases

(Adapted from ISO 14040 (2006a: 8), Figure 1, with the permission of ANSI on behalf of ISO, © ISO 2013, all rights reserved)

the scoping phase of an LCA. The inventory analysis quantifies emissions and the impact asessment translates emissions to impacts.

In order to interpret and apply the results of an LCA, the data must be assessed for quality and applicability. LCA results should only be applied to impact decisions about processes or purchasing after comprehensive interpretation of the LCA results. An LCA is refined as more information about the system is understood and the LCA results inform the analysis. Establishing the goal and scope as well as interpreting the LCA results define the analysis. The inventory and impact assessment phases contain the collection and assembly of data that informs both the refinement of the goal and scope definition and enables the interpretation of the results.

2.1 The LCA goal

ISO STANDARDS 14040 AND 14044 (ISO, 2006a, 2006b) require that the goal of an LCA study be unambiguously stated and include:

- the intended application (what);
- the reason for carrying out the study (why);
- the intended audience (for whom);
- whether the results are intended for comparative assertions intended to be disclosed to the public (how).

The goal of an LCA captures the questions about the environmental impacts of a process (e.g. manufacturing or waste treatment) or a product that inspires the analysis, and defines how the analysis will be used. LCA is often used to compare two or more options and evaluate if one is preferable. These LCAs are designed to inform decisions and are called a comparative assertion. ISO standards require additional analysis and care when performing the LCA used to make comparative assertions to help avoid inaccurate or misleading claims. See Chapter 5, LCA Standards, for more detail on these requirements. Examples of comparative LCAs in the building industry include:

- Which has a lower life cycle environmental impact: carpet tile or vinyl tile?
- Which window is preferable: a standard local window or a high performance window from overseas?
- Which structural system is preferable: steel, concrete or wood?

Life cycle assessments are also performed to quantify and qualify the environmental impacts of a specific building or building product. The results can help identify opportunities to reduce impacts, can target future research, and help policy-makers understand where to focus reduction efforts. Examples of building specific evaluation LCAs include:

- What are the total potential life cycle environmental impacts of a specific building?
- What are the environmental impacts related to manufacturing gypsum wallboard and what opportunities are there to reduce environmental impacts through changing manufacturing processes?
- What are the environmental impacts related to manufacturing a (GLB) glue laminated timber beam?

For example, a comprehensive goal statement for a glue laminated timber beam LCA could include the following:

WHAT – this LCA is a study of glue laminated timber beam manufacturing.

WHY – developed to understand the environmental impacts of the manufacturing process.

WHO – for use by manufacturers or structural engineers.

HOW – to integrate into a more comprehensive LCA.

BENEFIT – to manufacturers looking to understand and reduce the impacts of manufacturing.

LIMITS – the limitations of appropriate use of the LCA data (e.g. only for manufacturers' internal improvement or for publication for consumers to read).

LCA standards do not define what the appropriate goals are but rather require that an LCA goal be clearly stated. A formal critical review (see Section 5.1) of an LCA study can neither verify nor validate the goals but rather evaluates if the study conforms to the stated goals (ISO, 2006b). The user of the LCA data must evaluate the goal and the application of LCA results. The purpose of clearly defining an LCA goal is to help evaluators determine if the scope and detail of the study are adequate to achieve the stated goal(s). The goal of an LCA should be continually re-evaluated and improved as the analysis is developed and might be adjusted based on information discovered in the course of the analysis.

2.2 The LCA scope

THE SCOPE OF AN LCA requires the definition of what is included in and excluded from the analysis and defines the parameters of the study. ISO standard 14044 has specific requirements as to what must be included in the scope definition. The following is a summary of the key items that should be defined in an LCA scope as adapted from ISO (2006b: 7):

■   The product to be studied: Establishing function, performance and unit of analysis (functional unit).

■   The system boundary: What is excluded from and included in the analysis (which unit processes are part of the system studied).

■   Methodological choices: Including allocation assumptions and impact assessment and interpretation methods.

■   Analysis details: Sources of the data, data quality requirements and type of critical review (if any).

The LCA scope should support the goals of the LCA study. Depending upon the goal, different scopes of study will be appropriate. For example, if the goal of a study is to reduce water use, the scope of an LCA might be limited to tracking water and could ignore emissions to air. Or the impacts of water consumption might be ignored when the LCA goal is to focus on improving manufacturing energy efficiency. In some instances, the goal of an LCA might be redefined after gaining a better understanding of the feasible scope of the analysis, which may be practically limited, based on available data, time and expertise. Similar to the goal, the scope of an LCA should be continually re-evaluated as the analysis is developed. The life cycle stages typically included in building product and building LCAs have been standardized into four stages: product, construction, use and end of life, as detailed in Figure 2.2.

DEFINING THE PRODUCT to be studied is often not as simple as it might first appear, especially when comparing alternative options. An LCA typically identifies a unit of the product to analyze. For a manufactured product such as a disposable cup, the unit might be one cup. However, one could not compare a disposable cup to a re-usable cup as they are not functionally equivalent. The term "functional unit" defines a unit of analysis that includes quantity, quality and duration of the product or service provided.

A1: RAW MATERIAL SUPPLY

A2: TRANSPORT

A3: MANUFACTURING

A4: TRANSPORT

A5: CONSTRUCTION + INSTALLATION

B1: USE	B4: REPLACEMENT	B7: OPERATIONAL WATER USE
B2: MAINTENANCE	B5: REFURBISHMENT	
B3: REPAIR	B6: OPERATIONAL ENERGY USE	

C1: DEMOLITION	C4: DISPOSAL
C2: TRANSPORT	
C3: WASTE PROCESSING	

2.2 LCA scope in a typical building life cycle
(Adapted from CEN, 2011a: 14)

The functional unit of a coffee cup might then be to deliver 200ml (quantity) of 80° C liquid with insulation to be able to be held in hand (quality) up to twice in one day for ten years (duration). If an LCA study were based upon the assumption that 3,650 paper cups are used over ten years, the environmental impacts of one paper cup would be multiplied by 3,650 to compare with the equivalent single ceramic cup (provided one assumes that a ceramic cup can last that long).

In many instances, the analysis is performed based on a declared unit, which only includes quantity. Declared units are used when the precise function(s) of the product are not known or stated. Declared units are also used when creating LCI data for material manufacturing that can be used in many different applications (e.g. cubic meter of concrete that could be used as a column or a roadway).

CASE STUDY 2.1: FUNCTIONAL UNIT

LCA could be used to evaluate the environmental impacts of wood siding. If the analysis were performed for the declared unit (e.g. $1m^2$ of wood siding of a set thickness), the results would be useful for the manager of the plant to understand and reduce manufacturing impacts but not appropriate to use to compare to another siding product.

As shown in Figure 2.3, the functional unit of siding must address more than just quantity. A more careful functional unit definition would address the

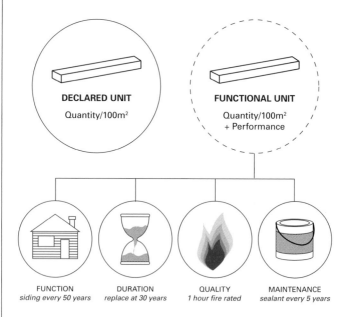

DECLARED UNIT
Quantity/100m²

FUNCTIONAL UNIT
Quantity/100m²
+ Performance

| FUNCTION | DURATION | QUALITY | MAINTENANCE |
| siding every 50 years | replace at 30 years | 1 hour fire rated | sealant every 5 years |

2.3 Declared vs. functional units

primary function (e.g. weather protection and aesthetics), performance (e.g. wind, fire and water standards, maintenance requirements) and time (e.g. over 50-year life span of a building) in addition to the quantity (e.g. area covered).

Performing a comparative LCA of two siding types to meet this function might identify that one material required replacement every 30 years and refinishing every five years while the other could last over 50 years with minimal servicing. With careful definition of the product function, the LCA is more likely to be a true "apples-with-apples" comparison. In some instances, some functions of the siding might not be taken into account (e.g. fire resistance) and the LCA report would need to explain and document this fact.

LCA results for two different products can be comparable if analyzing functionally equivalent systems. If looking to compare results of different analyses, ensuring that the function of both studies is equivalent is critical.

Table 2.1 outlines the functional and declared units for different building materials and products. For materials or products in which the application

Table 2.1 Declared and functional units of typical building products

Category	Declared Unit	Example of Functional Unit
Siding	1m²	Provide 1m² of exterior finish that conforms to typical residential aesthetic and provides UV light, water, fire and impact protection for 50 years.
Flooring	1m²	Provide 1m² of interior walking surface that conforms to typical residential aesthetic and provides slip resistance and impact protection for 50 years.
Concrete Wood Glass	1m³ at 3000 psi 1m³ stud grade 1m² at ½" thick	Not possible to define without context of application structure, sidewalk, finish, framing, installation method, orientation, etc.
Brick	1 brick	Non-load-bearing exterior finish conforming to typical commercial standards for UV light, water, fire and impact protection for 50 years.
Window	1m² double-gazed wood frame	Provide 1m² of window (including frame) for 30 years, meeting specified window performance standards with use phase impacts modeled.

is not known, an LCA based on functional units is not possible. As shown in this example, concrete could be used for a sidewalk, foundation or structural slab. Each of these end products would have distinct functional units. When an LCA is focused on the manufacturing stage of an intermediate product such as concrete, LCAs are often performed per a declared unit as the study can provide useful results without knowing the final end use of the product. See standards for EPDs of building products (ISO, 2007; CEN, 2011a) for more detailed requirements related to the use of functional and declared units.

The system boundary

An important consideration within the scope phase is to ensure that the product system of the study is clearly defined (what is to be studied and which components of the system are included). LCA requires a clear definition of the system boundaries (which industrial and natural processes are to be included and excluded). While the purpose of an LCA study is theoretically a comprehensive evaluation, in reality, the preparation of an LCA requires that choices be made about which processes to include in the study, which inputs and output to measure or estimate, which environmental impacts to evaluate and which data sources to use. The scope of the study should be defined in order to achieve the stated goal.

Often LCA studies will exclude the following processes:

- manufacture of fixed equipment (the factory);
- manufacture of mobile equipment (trucks);
- hygiene-related water use (toilets and sinks for workers);
- employee labor (commuting).

Some whole building LCA studies exclude processes such as construction energy use or mechanical and electrical systems. When known LCA stages are excluded, the goal of the LCA must be carefully assessed to ensure that the limited scope is still adequate to meet the requirements of the goal. Evaluating whether it is appropriate or not to omit these components requires an understanding of the goal of the LCA study and an appreciation of the significance of the omission. Often, the latter is not fully evaluated due to lack of resources.

When developing an analysis to compare materials, products or systems, defining the system boundary to effectively capture comparable components is critical to obtaining meaningful results. A cradle-to-gate study often is evaluated based on a declared unit (e.g. a cubic yard of concrete with defined compressive strength). A cradle-to-grave study used for comparisons between products or options is required to define a functional unit that would include quantity and performance but also critical life cycle performance characteristics such as duration/life span.

A system boundary diagram is a graphic representation of the scope of an LCA. An annotated system boundary for a generic process is shown in Figure 2.4. The system boundary diagram should clearly indicate what is included in the LCA analysis and which inputs and outputs have been tracked.

The system boundary must be defined to support the stated goal and scope of the LCA. As noted earlier, some LCAs only analyze impacts from cradle-to-gate while others include the whole life cycle. Other LCAs might focus on improving selected fabrication processes (e.g. welding, die-casting and extrusion) and thus only analyze from gate-to-gate. Gate-to-gate LCI data are typically found in LCI databases to be used by LCA practitioners.

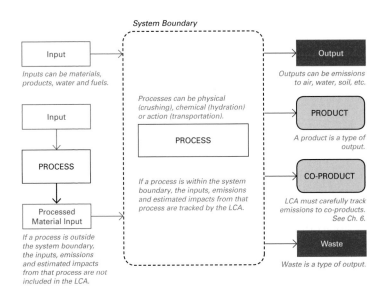

2.4 System boundary of a generic process

In other situations the goal of the LCA might be narrowly defined, for example, looking to reduce the environmental impact of finishing a wood product, thus all manufacturing steps not related to finishing could be excluded from the LCA scope while still achieving the goal of the analysis.

CASE STUDY 2.2: GLUE LAMINATED BEAM SYSTEM BOUNDARY

Figure 2.5 is a system boundary diagram describing an analysis of glue laminated beam manufacturing. In this cradle-to-gate analysis, the goal was to develop LCI for a declared unit of $1m^3$ of glue laminated timber. All the life cycle stages beyond the manufacturing stage are excluded. Additionally, the facilities' office operations and the factory infrastructure are excluded, which is common in LCA practice (given the assumption that the impacts are not relevant to product manufacturing or are relatively quite small on a per product basis.) Again, the validity of this assumption should be, but is not always, verified.

In this case, it is important to note that the biogenic carbon (carbon dioxide absorbed by the tree due to the biological growth of the tree) is not included in this analysis. This is generally considered a conservative assumption as it does not take into account the time-based "sequestration" of the bio-based carbon in long-life wood products used in buildings or address the fact that wood can remain inert for a significant period of time in a landfill. See Section 6.3 for a more detailed description of the intricacies of biogenic carbon, bio-fuels and carbon sequestration.

CASE STUDY 2.3: OFFICE BUILDING SYSTEM BOUNDARY

The system boundary for a cradle-to-grave LCA of a hypothetical commercial office building is shown in Figure 2.6. In this example, the furniture, office supplies, site work and worker transportation are excluded from the system boundary, as the goal of the study was to understand the relative environmental impacts due to construction, operation and end of life. In order to complete the LCA, the impacts of materials and construction are modeled based on rough estimates of material quantities; operational impacts are based on actual measured data; and end-of-life impacts are estimated based on average data and the typical regional practice of landfilling waste.

The inventory phase is described in Section 2.3. For a whole building LCA, the inventory phase would require quantification of the amount and type

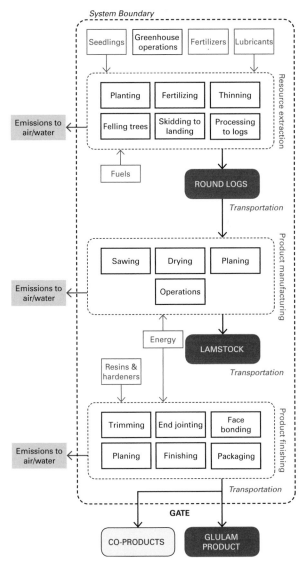

2.5 System boundary of glue laminated timber

(Adapted from Puettmann *et al.*, 2013)

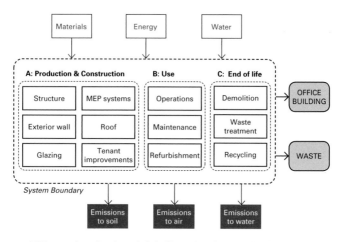

2.6 LCA system boundary for a whole building LCA of an example office building

of materials and energy used for each of the life cycle stages shown in Figure 2.6. These inventories can be created from primary data (the actual quantities of materials and energy used), calculated or estimated by industry experts and/ or LCA practitioners. Building industry-specific LCA tools can help ease the assembly of the quantity data and link to LCI databases to quantify the inputs and outputs and evaluate the potential environmental emissions and impacts for the building system studied.

Methodical choices

As part of clarifying the scope of the LCA, the choices related to LCA methods must be transparently communicated. LCA standards permit a wide range of methods to be employed provided that the analysis clearly documents the methods used for the study. Three key aspects that must be defined include:

1   allocation (see Sections 6.1, 6.2 and 6.3);
2   impact assessment metrics and methods (Chapter 3 and Section 4.4);
3   interpretations methods (Section 2.5).

Analysis details

The details of the analysis must also be defined. This should include a description of the following items that are defined in greater detail in Chapter 5: LCA Standards:

- source of data: measured, estimated, etc. (Chapter 4);
- data quality metrics (Sections 4.5 and 6.4);
- critical review (Section 5.1).

As with the methodical choices, LCA standards permit a wide range of analysis methods, data sources and assumptions. The standards require that these details be clearly defined within the LCA report to enable those reading and interpreting the data to understand the strengths and weaknesses of the study.

2.3 Inventory analysis

THERE ARE TWO KEY ASPECTS of LCI analysis: (1) collecting data (can be measured, calculated or estimated); and (2) calculating data to attain results for the system being studied.

ISO 14044 (2006b: 12) recommends classifying the inventory data into four categories: (1) inputs (energy, materials etc.; (2) products (including co-products and waste); (3) emissions (to air, water and soil); and (4) other environmental aspects as represented in Figure 2.7.

The results of an LCI report a compilation of quantities of resources consumed (from nature) and emissions (to nature) for a specific process. An LCI analysis can require a significant amount of data collection. The quantities of materials and energy used in each process and life cycle stage must be quantified (e.g. the amount of round logs and resin for glue laminated beam manufacturing and kilowatts of electricity consumed). Additionally, the emissions related to each process must be quantified (e.g. kg carbon dioxide released per kilowatt of electricity generated).

Life cycle inventory databases report LCI data unit process for a unit process (e.g. kg material produced) or for an aggregated process combining the many unit processes included in a system (e.g. cradle-to-gate production of a steel wide flange beam). Average inventories for specific manufacturing processes are typically created based upon surveys of material and energy use

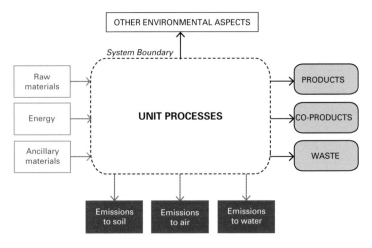

2.7 Life cycle inventory data classification

of manufacturers in a specific region and can report emissions based on estimates from computational models or actual measurements.

The result of an inventory analysis is a list of the total inputs from nature (raw materials, energy and water, etc.) and emissions to air, water and land for the system under study. See Chapter 5 for more detail on LCI databases. This list can include hundreds of different chemical emissions, which can be difficult for non-LCA experts to interpret.

CASE STUDY 2.4: LCI OF GLUE LAMINATED BEAM

Table 2.2 shows a partial LCI of glue laminated timber (GLB) production in the Pacific Northwest of the United States that matches the system boundary defined in Figure 2.3. This recently published data (Puettmann *et al.*, 2013) includes detailed descriptions of the data sources and methods and is intended to be used to update the US Life Cycle Inventory Database (NREL, 2013). Surveys were performed to represent the cross-section of forest management and manufacturing practices in western Washington and Oregon in 1999–2000 and have been updated with more current forestry operations, boiler, resin and electrical grid data.

Table 2.2 Partial life cycle inventory emissions to water and air in GLB manufacture: kg per m³ of product

Substance	Emissions (kg/m³)		
	To Air	From Forestry	GLB Production
Carbon dioxide, fossil	154.0000	9.4600	144.0000
Carbon dioxide, biogenic	145.0000	0.0075	145.0000
Sulfur dioxide	1.0600	0.0062	1.0500
Nitrogen oxides	0.7460	0.1710	0.5740
Particulates	0.5550	0.0009	0.5540
Methane	0.4830	0.0128	0.4700
Carbon monoxide	0.4530	0.0128	0.4530
Particulates, 2.5-10 µm	0.3870	0.0053	0.3810
Carbon monoxide, fossil	0.3460	0.0860	0.2600
VOC	0.3420	0.0046	0.3380
Particulates <2.5 µm	0.3170	0.0000	0.3170
2-Propoanol	0.1740	0.0000	0.1740
Particulates <10 µm	0.0844	0.0000	0.0844
Methane, fossil	0.0823	0.0010	0.0814

µm = micrometer

(Puettmann et al., 2013)

2.4 Impact assessment

RELATING RESOURCE USE AND ENVIRONMENTAL emissions to estimations of natural resource, natural environment and human health impacts is done in the impact assessment stage of an LCA. Depending upon the process, hundreds or even thousands of different emissions may occur and be reported by an inventory analysis. In order to translate emissions to impacts one must: identify which environmental impact categories to study, establish which emissions influence which impacts and calculate the total relative impacts (ISO, 2006b: 16). The order of the steps is important. The selection of environmental impacts to be studied should be done prior to knowing results to avoid the temptation of selecting impacts to study based on preliminary results (ILCD, 2010a: 108).

Environmental impacts are typically influenced by multiple different emissions and resources. Similarly a single emission might contribute to more than one environmental impact. For example, multiple emissions (including carbon dioxide and methane) have been determined to be greenhouse gases (GHGs) and impact the Earth's climate as described in Section 3.2. Impact assessment evaluates the relative contribution of multiple GHGs to climate change: one kg of methane has 25 times the impact of a kg of carbon dioxide. Methane emissions impact smog formation but carbon dioxide emissions do not.

The translation of emissions to impacts is computed through multiplying each emission by a characterization factor that represents the relative contribution of each emission to the environmental impact. Table 2.3 demonstrates the basic calculation method of chemical emissions. Emissions that do not have an impact on climate change, such as carbon monoxide, have a characterization factor of zero.

Table 2.3 Calculating mid-point impacts with characterization factors

Substance	Emissions (kg)		Characterization Factor (kg CO_2e/kg)		Impact (kg CO_2e)
Carbon dioxide	1.96	X	1	=	1.960
Methane	0.0005	X	25	=	0.0125
Carbon monoxide	0.00013	X	0 (not a GHG)	=	0
Nitrous oxides	0.0016	X	298	=	0.477

TOTAL CLIMATE CHANGE ESTIMATED IMPACT: 2.45 kg CO_2e

CASE STUDY 2.5: OZONE DEPLETION MID-POINT INDICATOR

Ozone depletion potential is another mid-point impact category that represents the broad range of human and environmental system damage caused by the depletion of stratospheric ozone. Figure 2.8 demonstrates the relationship of the mid-point impact category of ozone depletion to the final end-point categories that result in physical impacts to human and ecosystem health.

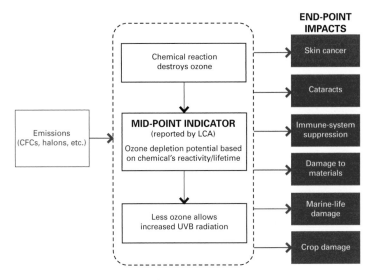

2.8 Ozone depletion mid-point and end-point impacts
(Adapted from Bare *et al.*, 2003: 54)

CASE STUDY 2.6: CALCULATING ENVIRONMENTAL IMPACTS

Organizations such as the US EPA (2012) and the University of Leiden (CML, 2002) have developed models that enable the computation of mid-point impacts through the multiplication of individual emissions by characterization factors that represent the relative impact of different emissions.

Table 2.4 shows the multiplication of the air emissions reported in the LCI of glue laminated timbers (Puettmann *et al.*, 2013) by the TRACI V2.1 (EPA, 2012) characterization factors for climate change impacts. Note the carbon dioxide emissions from fossil fuels are over 90 per cent of the total climate change impact, with methane constituting another 7 per cent. Note that the carbon dioxide emissions from biogenic fuels (combustion of wood chips, etc.) are nearly equivalent in mass to the fossil fuel emissions yet are not included in this chart as they are considered carbon-neutral (see Section 6.3).

Table 2.4 Climate change impacts for glue laminated timber manufacturing

Substance	Air Emissions kg	TRACI Factors	Impact $kgCO_2e$
Carbon dioxide, fossil	154.0000	1.00	154.00
Carbon dioxide, biogenic	145.0000	*	*
Sulfur dioxide	1.0600	0.00	0.00
Nitrogen oxides	0.7460	0.00	0.00
Particulates	0.5550	0.00	0.00
Methane	0.4830	25.00	12.08
Carbon monoxide	0.4530	0.00	0.00
Particulates, 2.5-10 µm	0.3870	0.00	0.00
Carbon monoxide, fossil	0.3460	0.00	0.00
VOC	0.3420	0.00	0.00
Particulates <2.5 µm	0.3170	0.00	0.00
Propoanol	0.1740	0.00	0.00
Particulates <10 µm	0.0844	0.00	0.00
Methane, fossil	0.0823	25.00	2.06
Carbon dioxide, other	0.0733	1.00	0.07
Sulfur oxides	0.0669	0.00	0.00
Ammonia	0.0027	0.00	0.00
Other	0.0000	0.00	1.64
Subtotal without biogenic carbon:			169.85

* not reported as carbon emissions related to bio-fuel combustion offset by carbon absorbed during growth

Correlating an emission to a specific environmental impact requires an understanding of the underlying mechanisms and the ability to model what happens between emission and impact. The end condition (fate) and transport (how and where it goes) are critical to understanding the environmental impact of an emission as well as the interaction. Chapter 3 outlines environmental impacts typically evaluated by an LCA.

Ideally, the reported impacts would enable an evaluation of the environmental, economic and social impacts associated with the process studied by the LCA. Methods to characterize economic and social impacts are less developed and thus not currently included in most LCA reports. Integrating LCA and life cycle costing has the potential to include economic impacts. Social impacts are particularly difficult to quantify and agree upon, given the wide range of values related to measures of social success or damage.

While LCA standards require that "all" relevant environmental impacts be assessed, in practice, some impacts are difficult to capture with conventional LCA studies.

CASE STUDY 2.7: ACIDIFICATION IMPACTS

Figure 2.9 represents a simplified explanation of transport and fate for acidification impacts. Chemical emissions (including sulfur dioxides and nitrogen oxides) from sources such as a coal power plant are released into the air. These chemicals are transported in the air depending on wind and climate conditions (transport) and often carried to ground or water bodies via rain (fate). The final, or "end-point" environmental impact of acid rain includes a broad range of impacts, ranging from corrosion of buildings to lack of productivity of soils.

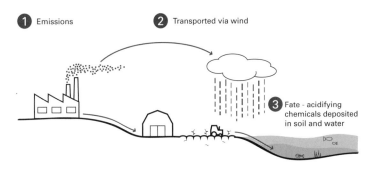

1 Emissions   2 Transported via wind

3 Fate - acidifying chemicals deposited in soil and water

2.9 Fate and transport

2.5 Interpretation

CRITICAL THINKING AND CAREFUL EVALUATION of the strengths and weaknesses of the LCA are essential during the interpretation phase in order for the conclusions, limitations and recommendations resulting from an LCA to be meaningful.

Figure 2.10 demonstrates the key components of interpretation as identified by ISO 14044. Simultaneously with interpreting LCA results, the goal and scope should also be assessed, and potentially revised, to evaluate if they are appropriate given the resulting data.

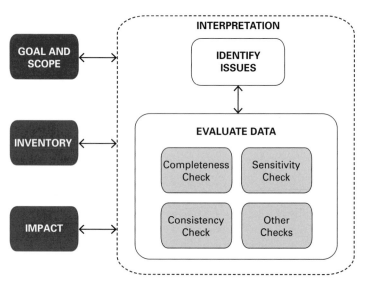

2.10 LCA interpretation

(Taken from ISO 14044 (2006b: 24), Figure 4, with the permission of ANSI on behalf of ISO, © ISO 2013, all rights reserved)

After computing the LCI and impact assessment, the data should be analyzed to determine if significant issues have been identified. ISO (2006b: 26) requires a minimum of three evaluation methods to be summarized and adapted, as shown below:

- **Completeness check:** Evaluating the LCA information and data to ensure that it is complete: that there are no missing emissions to environment. If items are found to be missing, new data and analysis are required. Alternately, the LCA goal and scope can be modified if the data is not available or time/money limitations preclude refining the study, and this limitation should be included in the final published report.

- **Sensitivity check:** Evaluating how sensitive the analysis is to changing assumptions or choices to assess the reliability of the final results and conclusions. This might lead to the need to obtain better data in order to understand key aspects of the LCA.

- **Consistency check:** Evaluating the study to ensure the analysis is internally consistent and matches the established goal and scope.

Additional data evaluation methods could include: quantifying the variability of both the LCI input data and the processes being analyzed, contextualizing the results to help users understand the significance (e.g. percentage of national consumption), and evaluating which processes contribute the largest share to select environmental impacts.

Conclusions, limitations and recommendations based on an LCA study are only appropriate after the results have been evaluated for significance and the data has been assessed. Sometimes, the conclusion may be that the initial stages of the LCA need to be reconfigured or redone. On other occasions, the evaluation of data quality issues is reported within the final results of the LCA.

CASE STUDY 2.8: REFINING THE GOAL AND SCOPE

An ambitious architect might start with the goal of creating a comprehensive cradle-to-grave LCA of the building being designed, as outlined in Figure 2.6. After completing the first iteration of the LCA and interpreting the results, the following inconsistencies were found:

- unknown material quantities for electrical system;
- unknown material types for plumbing systems;
- unknown demolition energy;
- tenant improvement impacts small and uncertain.

In order to finalize a complete LCA that conforms with ISO requirements, the scope was modified to eliminate MEP, tenant improvements and demolition energy from the system boundary as shown in Figure 2.11. Additionally, the goal could be changed from "determining a comprehensive understanding of the full life cycle impacts of the building" to "identifying components with relatively significant contributions to the total environmental impact of the building".

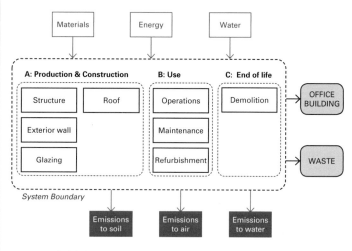

2.11 Revised whole building LCA system boundary

chapter 3

Environmental Impacts

ENVIRONMENTAL LIFE CYCLE ASSESSMENT is typically performed to understand and reduce environmental impacts. LCA tracks the extractions from and emissions to nature. As introduced in Section 2.4, Impact Assessment, methods have been developed to quantify the relative impact of emissions on different aspects of the environment. For example, GHG emissions are reported as climate change impacts, an impact mid-point that aggregates all emissions that are assessed to contribute to that impact in proportion to their relative influence on climate change.

As stated earlier, an LCA would ideally report "all" environmental impacts. However, due to limitations of available data, lack of consensus in modeling methods and the need to limit the time and money spent on an LCA, most LCAs track and report between five and eight environmental impacts and the inventory data (actual quantity of material/resource consumed) of one to more than ten categories. Consensus upon the best methods to characterize potential environmental impact is still developing (Hauschild *et al.*, 2012). See Table 3.1 for a summary of common impacts and Table 3.2 for inventory items commonly reported in LCAs of buildings and building products.

LCA standards (ISO, 2007 and CEN, 2011a) for building products have outlined the impacts and inventories that are required when reporting LCA results as an Environmental Product Declaration, or EPD, (see Section 5.3). Some impacts, such as climate change/GHG emissions, can be reported with clarity, as they have a high degree of global agreement on reporting methods (the US and European Union methods of reporting GHG emissions are nearly identical, the US requires tracking and reporting one more GHG not required by EU standards). Other impacts, such as human health or habitat disruption, have less developed methodology and there is a lack of clear consensus on how to compute and report them. Resource uses, such as water use or waste creation, are reported based on an absolute quantity consumed

Table 3.1 Environmental impacts commonly estimated by LCA

Environmental Impact	Units
Acidification	kg-SO_{2e}
Climate Change/Global Warming	kg-CO_{2e}
Eutrophication	kg/Ne
Ozone Depletion	kg CFC-11$_e$
Human Health	varies
Photochemical Ozone Creation/Smog	kg Ethane
Depletion of Abiotic Resources (elements)	kg Sg$_e$
Depletion of Abiotic Resources (fossil)	MJ
Depletion of Non-Renewable Energy Resources	MJ
Depletion of Non-Renewable Material Resources	MJ

Table 3.2 Resource use commonly estimated by LCA

Resource	Units
Use of Renewable Material Resources	MJ
Use of Renewable Primary Energy	MJ
Consumption of Freshwater	M^3 or liters
Use of Secondary Material	kg
Hazardous Waste Produced/Disposed	kg
Non-Hazardous Waste Produced/Disposed	kg
Radioactive Waste Disposed	kg
Components for Re-Use	kg
Materials for Recycling	kg
Materials for Energy Recovery	kg
Exported Energy per Energy Carrier	MJ

or released. Detailed standards to clarify and classify consumption are still being refined. LCA reporting of these inventory amounts requires the user to evaluate the relative environmental importance. Generally speaking, newer LCA standards tend to require reporting of greater range of impacts and in greater levels of detail.

Interestingly, emissions and impacts are quite interconnected. Combustion of fossil fuels impacts acidification, climate change, eutrophication and ozone production. Carbon dioxide causes both climate change and acidification. Eutrophication results in increased decomposition of organic matter that reduces oxygen and increases acidification. At the highest level, we are looking to protect human and ecosystem health. End-point measures of human health would include metrics such as increased death due to cancer and increase in incidence of asthma. End-point measures of ecosystem health would include metrics such as changes in species diversity and depletion of resources. As discussed in Chapter 2, LCA reports mid-point indicators representing relative emissions. See Figure 3.1, which demonstrates the relevance of typical LCA mid-point impacts to human and ecosystem health.

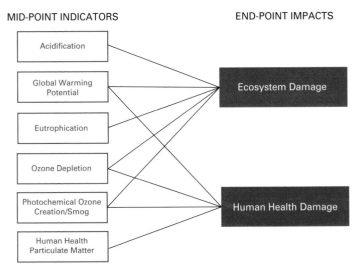

3.1 LCA impacts and relevance to human and ecosystem damage

Environmental impacts vary in their site specificity, as they are global in influence and impact or regionally varied. The emission of GHGs anywhere in the world results in the same global impact on climate change. Therefore, tracking where GHG emissions are released is not required in order to understand the global environmental impacts. The results of other emissions are highly dependent upon where they are released (see Figure 3.2). Impacts such as smog or acidification are highly dependent on local weather patterns and geography to influence the transportation of emissions (transport) and final destination of the emissions (fate).

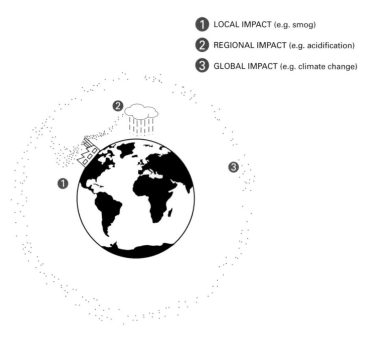

① LOCAL IMPACT (e.g. smog)

② REGIONAL IMPACT (e.g. acidification)

③ GLOBAL IMPACT (e.g. climate change)

3.2 Local vs. global relationship between emission and impact

The risk that emissions will result in environmental harm varies depending on regional ecosystem conditions. The US Environmental Protection Agency (EPA) reports impact characterization factors with its Tool for the Reduction and Assessment of Chemical and Other Environmental Impacts

(TRACI) (EPA, 2012; Bare *et al.*, 2003) with aggregated US risks. A more detailed impact assessment would address the differences between regions as varied as the desert south-west and the industrialized north-east. Europe has developed characterization factors for impacts such as eutrophication and acidification, reflecting European environmental variation.

In order to understand the risk of depositing these regionally relevant pollutants, one must understand if the region currently is (or is expected to have) exceeded the critical load. The concept of critical load, or the exposure "below which significant harmful effects ... do not occur" (UNECE, 2012: 4), is used to evaluate regionally varied impacts such as acidification and eutrophication.

What follows is a brief description of the most common environmental impact results reported in LCAs including information about which emissions and mechanisms contribute to the specific environmental degradation.

3.1 Acidification

ACIDIFICATION IS THE PROCESS OF CHANGING the pH balance of water and soil, resulting in more acidic or more basic substances. Plants and animals have optimum pH levels within which they thrive and a wider limiting range within which they can live. Ocean acidification and acidification of soils and water are similar but different in their mechanisms and effect.

A change in the pH balance reduces the productivity of soil, and in water can lead to a decline in plant and animal health. Some marine species such as plankton, oyster and sea urchins have demonstrated significant sensitivity to changes in the acid balance (Doney *et al.*, 2009). These organisms are a key part of the food chain and thus their decline could have significant impact on the overall vitality of many other species (Fabry *et al.*, 2008), including people whose nutrition and livelihood depend on the sea. The emissions that cause acidification can damage exposed building materials and finishes, leading to reduced durability, and increased maintenance of facilities.

One of several chemical reactions related to acidification is the combination of carbon dioxide, CO_2, with water, H_2O, to form carbonic acid, H_2CO_3 (Doney *et al.*, 2009). Ocean acidification is impacted by human activities such as carbon dioxide production as shown in Figure 3.3.

The generation of electricity through fossil fuel combustion is a prime source of CO_2, and to a lesser extent a contributor to sulfur and other gas

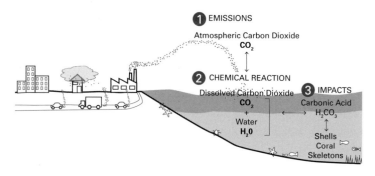

3.3 Ocean acidification process

emissions that result in man-made acidification. Chemicals that contribute to acidification are sulfur oxides (SO_x), which includes SO, SO_2 and SO_3, nitrous oxides (NO_x) and ammonia (NH_3). These chemicals can travel long distances as airborne gases before being deposited on the ground or in water as dust or soot or through rain, fog or snow, known as "acid rain" (EPA, 2013a) (Figure 3.4). Natural sources, including volcanic eruption and decomposition of plants, can also lead to acidification. Figure 3.5 outlines the primary contributions to acidification by industrial sector for Europe (EEA, 2012). Note that ammonia production is dominated by agricultural practices while energy production and transportation dominate SO_x and NO_x emissions. Buildings' impact on acidification is thus primarily due to the SO_x and NO_x emissions related to electrical use in operating buildings, electrical use in manufacturing and transportation of materials.

LCA acidification impacts are typically reported as emissions of equivalent sulfur dioxide (SO_{2e}) emissions. Often the SO_{2e} emissions to air and water are added together to report a single impact value. While each emission that contributes to acidification is scaled to the equivalent impact of one kg of SO_{2e}, fine-tuned regional differences are not typically accounted for. The effect of acidifying emissions depends upon the region: the weather conditions that impact the transport as well as the existing acid balance of the soils and water where the emissions are finally deposited (Bare, 2011).

Acid rain is a localized effect with impacts often seen directly downwind of the source. Estuaries, lakes and other water bodies near humans can

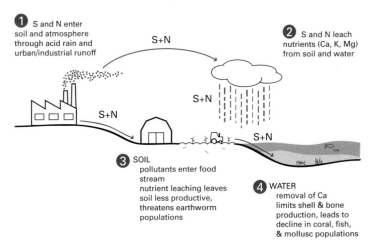

1 S and N enter soil and atmosphere through acid rain and urban/industrial runoff

S+N

2 S and N leach nutrients (Ca, K, Mg) from soil and water

S+N

S+N

S+N

3 SOIL
pollutants enter food stream
nutrient leaching leaves soil less productive, threatens earthworm populations

4 WATER
removal of Ca limits shell & bone production, leads to decline in coral, fish, & mollusc populations

3.4 Acidification of soil and water

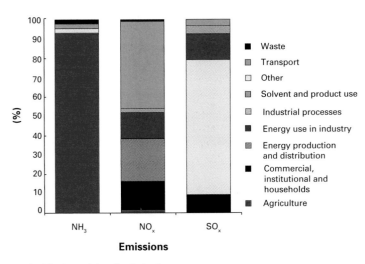

- Waste
- Transport
- Other
- Solvent and product use
- Industrial processes
- Energy use in industry
- Energy production and distribution
- Commercial, institutional and households
- Agriculture

Emissions

NH_3 NO_x SO_x

(%)

3.5 Acidification emissions distribution by sector
(EEA, 2012)

also see local effects from runoff or dumping. CO_2 gases remain in the atmosphere and have a global impact on ocean acidification.

In order to reduce the acidification impact of building materials and products, the primary focus should be reducing the combustion of fossil fuels and selecting fuels and systems that reduce the SO_X and NO_X emissions, such as switching from coal to natural gas or improving vehicle emissions. Additionally, reducing carbon emissions from combustion and manufacturing processes such as cement production will decrease acidification.

3.2 Climate change

CLIMATE CHANGE INCLUDES THE IMPACTS of global warming, the increase in the global average temperature of the Earth's surface, as well as other significant changes to climate such as precipitation or wind that occur over several decades or longer (EPA, 2013b). LCA characterizes GHG's potency as compared to an equivalent mass of carbon dioxide (CO_2). In addition to

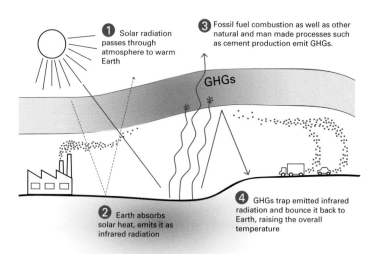

3.6 Climate impacts from GHG emissions

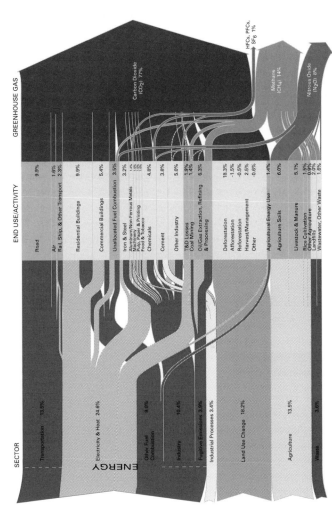

SECTOR

Transportation 13.5%

Electricity & Heat 24.6%

Other Fuel Combustion 9.0%

Industry 10.4%

Fugitive Emissions 3.9%

ENERGY

Industrial Processes 3.4%

Land Use Change 18.2%

Agriculture 13.5%

Waste 3.6%

END USE/ACTIVITY

Road 9.9%
Air 1.6%
Rail, Ship, & Other Transport 2.3%

Residential Buildings 9.9%

Commercial Buildings 5.4%

Unallocated Fuel Combustion 3.5%

Iron & Steel 3.2%
Aluminum/Non-Ferrous Metals 1.4%
Machinery 1.0%
Pulp, Paper & Printing 1.0%
Food & Tobacco 1.0%
Chemicals 4.8%

Cement 3.8%

Other Industry 5.0%

T&D Losses 1.9%
Coal Mining 1.4%
Oil/Gas Extraction, Refining 6.3%
& Processing

Deforestation 18.3%
Afforestation -1.5%
Reforestation -0.5%
Harvest/Management 2.5%
Other -0.6%

Agricultural Energy Use 1.4%

Agriculture Soils 6.0%

Livestock & Manure 5.1%

Rice Cultivation 1.5%
Other Agriculture 0.9%
Landfills 2.0%
Wastewater, Other Waste 1.6%

GREENHOUSE GAS

Carbon Dioxide (CO2) 77%

HFCs, PFCs, SF6 1%

Methane (CH4) 14%

Nitrous Oxide (N2O) 8%

3.7 Global distribution of energy use and GHG emissions
(Adapted from WRI, 2005)

CO_2, gases such as water vapour (H_2O) and methane (CH_4) and nearly 100 other documented gases work to impact climate. This impact is reported in LCAs in units of CO_{2e} as the Global Warming Potential or GWP, or GHG emissions, or a "carbon" footprint, or climate change impact.

The GHGs absorb energy, trapping the heat close to the Earth, functioning like a physical greenhouse trapping heat within a glass building (Figure 3.6 see p. 44). GHG emissions originate from a variety of sources. In the building industry, emissions from energy generation, both for the operation of buildings and production of materials used to construct buildings, are a critical component. Some industrial processes, such as production of cement, release CO_2 as a result of chemical reactions within the manufacturing processes. Biological processes such as plant growth (which absorbs carbon) and decay (which releases carbon), and agricultural production (methane and nitrous oxide emissions) also contribute significantly to global GHG emissions.

A visual representation of GHG emissions by sector adapted from data compiled by the World Resources Institute is shown in Figure 3.7 (see p. 45). In order to understand the full scope of the impact of the building sector, the emissions from industry and transportation related to manufacturing building products and materials, transporting them and maintaining and demolishing buildings should also be included. See Section 6.3 for more detailed information on tracking and reporting the carbon absorbed, sequestered and released throughout the life of wood and other bio-based building products.

3.3 Eutrophication

EUTROPHICATION IS THE EXCESS OF BIOLOGICAL ACTIVITY in an aquatic system, typically resulting from the addition of nutrients, such as nitrogen, to water bodies, and it can also be called nutrient pollution. The excessive plant growth, in turn, strangles the water system by depleting the available oxygen needed for life underwater. Algae blooms and dead zones in water are common manifestations of the problems of eutrophication (EPA, 2013h).

Urban sources of eutrophication include septic field seepage, storm and wastewater runoff – particularly from fertilized landscapes, and fossil fuel combustion. Rural sources of eutrophication include rainwater runoff after contact with fertilized agriculture and manure and aquaculture. See Figure 3.8 for a representation of eutrophication mechanisms.

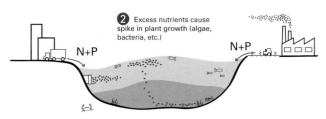

1 Agricultural, urban, and industrial runoff adds excess nutrients (N + P) to water systems

2 Excess nutrients cause spike in plant growth (algae, bacteria, etc.)

N+P   N+P

3 New plant population consumes all available oxygen, creating dead (hypoxic) zones and causing other species to flee or die

3.8 Eutrophication

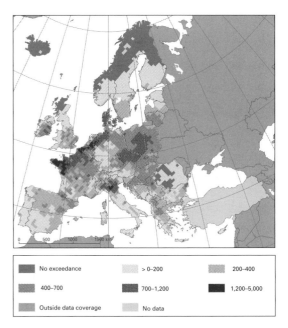

No exceedance	> 0–200	200–400
400–700	700–1,200	1,200–5,000
Outside data coverage	No data	

3.9 European exceedance of critical loads for eutrophication due to the deposition of nitrogen in 2010

(EEA, 2012)

Eutrophication risk has a large regional variability. As discussed in Section 3.1 on acidification, impacts are most significant if emissions are released in a region in which the critical load is close to or already exceeded. Figure 3.9 (see p. 47) is a map of European regions in which the amount of nitrogen exceeds critical loads (EEA, 2012), demonstrating that emissions of nitrogen in areas such as the Netherlands would be more critical than in northern Sweden. Figure 3.10 plots the magnitude of emissions per region. Note that a strong correlation between the two exists.

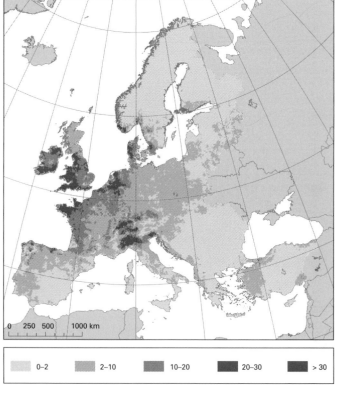

3.10 Annual European diffuse agricultural emissions of nitrogen to freshwater (kg nitrogen per hectare of total land area)

(EEA, 2012)

3.4 Ozone depletion

OZONE (O$_3$) IS A GAS THAT IS PRESENT in the atmosphere and in constant flux, undergoing chemical reactions with other elements such as chlorine and oxygen. Ozone high in the stratosphere functions to protect the Earth from the sun's ultraviolet (UV) rays. Ozone close to the Earth's surface is characterized as smog and its environmental impact is described in Section 3.5.

In the 1970s, scientists made the connection between decreasing levels of ozone discovered from satellite data (Krueger and Minzner, 1976) and the increase of chlorofluorocarbons (CFCs) in the atmosphere due to the chemical chain reaction caused by the chlorine released when CFCs degrade in the stratosphere (Molina and Rowland, 1974; Rowland and Molina, 1994). CFCs react with the energy of sunlight and release chlorine. The chlorine reacts with ozone in a cycle, breaking the ozone down into oxygen. The chemical equation can be reduced to the transformation of a single Oxygen atom (O) plus Ozone (O$_3$) into two Oxygen molecules (2O$_2$) and is represented in Figure 3.11. Figure 3.12 shows the process of ozone depletion.

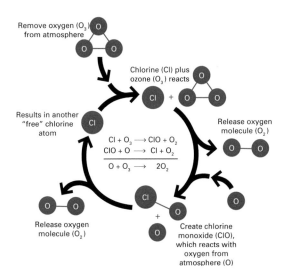

Remove oxygen (O$_3$) from atmosphere

Chlorine (Cl) plus ozone (O$_3$) reacts

Results in another "free" chlorine atom

Release oxygen molecule (O$_2$)

$$Cl + O_3 \longrightarrow ClO + O_2$$
$$ClO + O \longrightarrow Cl + O_2$$
$$\overline{O + O_3 \longrightarrow \quad 2O_2}$$

Release oxygen molecule (O$_2$)

Create chlorine monoxide (ClO), which reacts with oxygen from atmosphere (O)

3.11 Chemical reaction resulting in ozone depletion

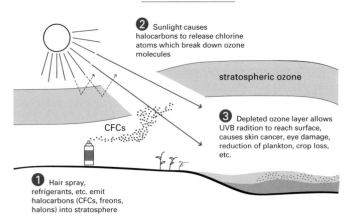

3.12 Ozone depletion process and impacts

Impressive global cooperation focused on reducing the emissions of CFCs occurred which is summarized in Twenty Questions and Answers about the Ozone Layer: 2010 Update (Fahey and Hegglin, 2010). The development and adoption of the United Nations Montreal Protocol, which is the only UN treaty to achieve universal ratification (196 nations) was ratified in 1987. Figure 3.13 plots the CFC emissions over time, demonstrating how emissions have been radically reduced since the treaty was enacted.

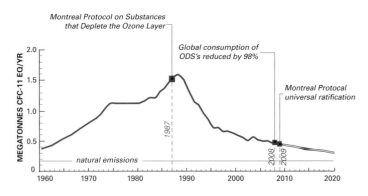

3.13 Success of the Montreal Protocol

(Adapted from Fahey and Hegglin, 2010: Q.2)

By 2010, 98 per cent of the historic levels of ozone-depleting substances had been phased out (UN, 2012). In the building industry this resulted in the changing of refrigerant chemicals and modifications to manufacturing processes.

3.5 Smog formation

THE FORMATION OF GROUND-LEVEL OZONE results in smog, and is a serious air quality problem. Ground-level ozone poses a significant threat to human health, causing irritation of respiratory systems, reduced lung function, aggravation of asthma, and may cause permanent lung damage. The impacts on children, active adults and people with respiratory diseases, such as asthma, are most severe (EPA, 1999).

Excess ozone in the lower stratosphere causes smog while in the upper stratosphere ozone protects against the sun's harmful rays. This dichotomy of "good" vs. "bad" ozone is represented in Figure 3.14. The major sources of

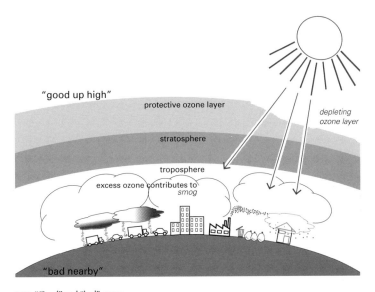

3.14 "Good" and "bad" ozone

ground-level ozone are the SO_x and NO_x emissions related to fossil fuel consumption and the release of volatile organic compounds (VOCs). Building materials, including paints and solvents, as well as building maintenance products are a significant source of VOCs. VOCs can also have human health impacts.

LCA typically reports smog formation in weight of ozone (O_3) emitted to air. Over 1,000 different chemical emissions have been attributed to smog formation (EPA, 2012). As is the case with acidification, the location of the emission and the local conditions are essential to understanding the environmental impact of smog formation. In regions such as Mexico City, with perimeter mountain ranges creating a natural "bowl" to trap low air, added emissions are more likely to exceed the critical level needed to cause negative health impacts as shown in Figure 3.15.

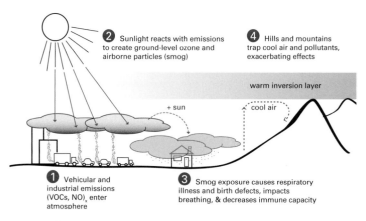

2 Sunlight reacts with emissions to create ground-level ozone and airborne particles (smog)

4 Hills and mountains trap cool air and pollutants, exacerbating effects

warm inversion layer

+ sun

cool air

1 Vehicular and industrial emissions (VOCs, NO)$_x$ enter atmosphere

3 Smog exposure causes respiratory illness and birth defects, impacts breathing, & decreases immune capacity

3.15 Ground-level ozone/smog formation processes

3.6 Human health and eco-toxicity

CHEMICAL AND (PM) PARTICULATE MATTER EMISSIONS can have wide-ranging impacts on human health, causing and aggravating diseases such as asthma, heart disease, low birth rate and cancer. As noted in Figure 3.1, other measured environmental impacts have a direct and indirect impact on human health. Correlating between emissions and human health requires integration of models that can predict the fate (where the emission ends up), that can predict exposure dosage to humans as well as model the relationship between dosage and human response (Figure 3.16). Environmental and biological scientists continue to advance methods to quantify these connections. The three most developed methods in LCA are: (1) reporting impacts to respiratory health related to quantities of particulates emitted; (2) evaluating the eco-toxicity of chemicals (Humbert *et al.*, 2011); and (3) declaring the chemical composition of materials or products.

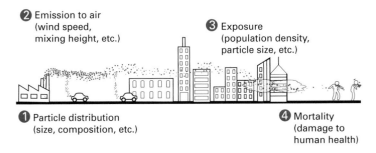

2 Emission to air (wind speed, mixing height, etc.)

3 Exposure (population density, particle size, etc.)

1 Particle distribution (size, composition, etc.)

4 Mortality (damage to human health)

3.16 Particulate matter emissions, transportation, fate, exposure and impacts

Respiratory health

Particulate matter (PM) is the mixture of very small particles and drops of liquids, including soil and dust particles as well as chemicals and metals (EPA, 2013f). PM is classified based on size. Small particles (less than 10 micrometers in diameter) can pass through our nasal passages. These particles cause serious health impacts when they get to our lungs. Particles are typically reported into two categories: (1) coarse (2.5–10 micrometers) and fine (less than 2.5 micrometers) (Figure 3.17).

LCA reporting of PM impacts can be done as a mid-point indicator (equivalent PM 2.5 emissions) as in the current version of the US EPA's TRACI V2.1 methodology (Bare, 2011). This accounts for transportation and exposure factors for different emissions but does not report the expected end-point impacts of disease and death.

Starting in 2005, a joint commission supported by the United Nations Environment Program (UNEP) and the Society for Environmental Toxicology

1 Coarse emissions from dusty roads and industries.

2 Fine emissions in smoke from fires, power plants, and auto exhaust.

3.17 Particulate matter emission sources

and Chemistry (SETAC) collaborated to develop harmonized models for reporting human and freshwater eco-toxicological impacts of chemicals in life cycle impact assessment. This model is called USEtox (Hauschild *et al.*, 2008). This program has produced consensus on the relative impacts of many common substances and reports human health impacts in units of comparative toxic units (CTUh), which is equal to disease cases per kg emitted. Cancer and non-cancer cases are added together.

Given the large uncertainty in predicting health impacts of chemical releases, another approach is to review the materials used in manufacturing and identify a list of "chemicals of concern". Product manufacturers can thus declare if their product includes any materials on this list. However, developing consensus on this list is difficult and as new chemicals and new knowledge of chemical risks continue to develop, there is no global agreement on which chemicals should be included in this list. Additional information related to issues such as toxicity and human health and a material content list that identifies materials of concern can be reported by an LCA. In practice, detailed material lists are not often provided.

Two not-for-profit organizations have developed methods in an attempt to advance material content transparency: the Living Building Challenge (ILFI, 2012), and the Health Product Declaration (HPD) Collaborative (HPD, 2012). The Living Building Challenge is an aspirational green building certification standard that has published a "Red List" of chemicals to be avoided in products used in their buildings. The HPD is an open source format for declaring content of product ingredients in a range of levels from fully transparent to limited identification of hazards. The HPD has compiled a list that they term the "Priority Hazard List" which compiles lists from government agencies such as the US EPA and the European Commission, as well as other sources. In order to comply with these transparency standards, manufacturers are required to review all these lists to identify potential risks from chemicals used in the manufacture of their products.

3.7 Resource use and depletion

THE CONCERN OVER THE DEPLETION of the Earth's resources (both material and energy), effectively introduced by *The Limits to Growth* (Meadows *et al.*, 1972), is highlighted by media focus on issues such as Peak Oil. Peak Oil is defined as the point in time where the production of crude oil grows to a maximum (the peak) and then gradually declines to zero (Bardi, 2009). Resource use and depletion are relevant to track by LCA, given the broad environmental, social and economic impacts of declining resources. LCA can evaluate resource use (reporting consumption, a current activity) or assess depletion (a prediction of future consequences) (Yellishetty *et al.*, 2011). Thus resource use can be reported within the LCA impacts (if integrating depletion) or the LCA inventory (if reporting quantities).

Non-renewable resources are those that do not renew themselves within a human time scale. Fossil fuels were generated by biologic mechanisms of the conversion of decomposing plants and animals to fossil fuels over millions of years. Yet the rate at which they are being created is nowhere near the rate at which we are extracting them. They are not considered to be renewable within the human time scale.

Resource consumption
As noted, the inventory results of an LCA can report consumption directly. Typically the use of material and energy resources are converted into units of energy, often termed "embodied" energy, representing both the energy available within a product (what could be attained if the product were incinerated) and the energy used as fuel to manufacture the product. The heat released during combustion of materials can be characterized by the gross calorific value (GCV or Higher Heating Value) or net calorific value (NCV or Lower Heating Value). Europe has typically reported results as NCV and the US has typically reported results as GCV.

As noted in Table 3.3, distinctions regarding the type of resource consumption can provide additional detail. For example, crude oil can be used in the creation of fuel to make energy or as a material input to the creation of plastic. Recognizing when resources are related to materials or energy helps to provide a finer-grained understanding of the type of resource consumption taking place. Energy resources might more easily be converted to renewable sources (e.g. solar power) than changing material resource consumption.

Table 3.3 Resource consumption impacts of LCA

	Type	Example	Units
Non-Renewable	Energy	coal combustion for energy	MJ
	Material	crude oil as material input to plastic	MJ
Renewable	Energy	bio-fuel	MJ
	Material	wood in home	MJ

Energy resources typically tracked as both energy and material resources from LCA are outlined in Table 3.3. Note there is no international standard to define what should be included in this calculation. As with many LCA methods, sound judgement is required by the LCA practitioner completing the study.

Resource depletion

Alternatively, or additionally, the relative impact of consumption of different resources can be characterized and reported as a weighted impact. This impact takes into account both the quantities of resources consumed and the impacts (environmental, economic and social) of their consumption. The impacts are typically reported for both abiotic elements (non-living resources such as minerals, chemicals and ores) and abiotic (fossil-based resources) as shown in Figure 3.18.

The University of Leiden's Institute of Environmental Sciences (CML) has published the most cited method reporting resource depletion as an environmental impact (van Oers et al., 2002). The Abiotic Depletion Potential (ADP-elements) reported by the CML method considers factors such as the ultimate reserve available, the resources that are potentially available, given technologic and economic constraints and the reserve that is currently available. Abiotic resources are tracked in units of an equivalent

	TYPE	EXAMPLE	UNITS
ABIOTIC DEPLETION	*Elements* *ADP-elements*	minerals, chemicals taking into account severity	kg-Sb$_e$
	Fossils *ADP-fossil*	crude oil use	MJ

3.18 Resource depletion impacts of LCA

reference material, Antimony (Sb), and reported in units of kg of Sb$_e$. Other calculation methods exist (Anderson and Thornback, 2012) and are in development.

Abiotic depletion of fossil fuel resources, ADP-fossil, is most commonly reported in energy values only, not accounting for the relative scarcity of different fuel types. Fuels are converted to net calorific or gross calorific values based on the relative heating values of different fuels. There can be a significant variation of the energy values of different fuel types and sources. LCI databases can contain a wide range of LCI data for different fuel types but do not necessarily have data for the exact fuel under consideration (e.g. coal from West Virginia vs. Germany). Therefore, again, LCA practitioners need to use their judgement in determining which dataset to use in assessing abiotic resources. Results prepared by different LCA practitioners will thus not necessarily be comparable unless both studies report using the same methods and using the same source data. Currently there is no consensus on the best source of data to use.

3.8 Consumption of freshwater

Water is a critical natural resource upon which all social and economic activities and ecosystem functions depend … There are major uncertainties about the amount of water required to meet demand for food, energy and other human uses, and to sustain ecosystems. These uncertainties are compounded by the impact of climate change on available water resources.

(UN 2012: 2)

WATER CONSUMPTION CAN BE TRACKED as an inventory item within an LCA or developed independently as a "water footprint" (Ercin and Hoekstra, 2012). In both of these methods the severity of the impact is not accounted for, as local conditions of scarcity are not included in the assessment. The interpretation of the results must thus integrate knowledge of regional water scarcity issues. There is a wide range of water resources available around the world and as one would expect, water consumption in Egypt has a significantly greater impact than water consumption in Norway.

Water footprints divide water consumption into blue (fresh surface or groundwater), green (precipitation on land that stays on surface/in soil) and grey (volume required to dilute pollutants) (Figure 3.19). Note the term grey water is also used in architecture to characterize the water reclaimed from sinks and other lightly used water sources in buildings to be re-used on site.

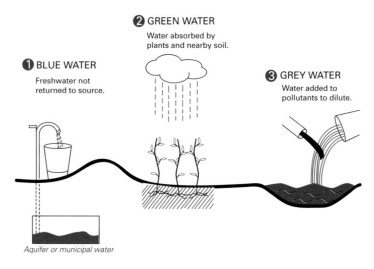

❷ GREEN WATER
Water absorbed by plants and nearby soil.

❶ BLUE WATER
Freshwater not returned to source.

❸ GREY WATER
Water added to pollutants to dilute.

Aquifer or municipal water

3.19 Water footprint classification in LCA

Typically LCAs report blue water consumption, though this distinction is not consistent or often not clearly documented. Water consumption during product manufacturing typically occurs in one of the following four mechanisms (Hoekstra *et al.*, 2011: 24):

1   water evaporates;

2   water is incorporated into the product;

3   water is not returned to the source;

4   water does not return in the same period (e.g. withdrawn in dry season and returned in wet season).

Water use attributed to building products typically includes processes such as: irrigation for bio-based products, groundwater or utility water use during the manufacturing process and released to the city wastewater treatment, and water incorporated into composite materials such as concrete. Water use can be measured, modeled or estimated. Models that integrate site-specific issues such as precipitation and soil permeability must estimate green water quantities.

A significant volume of water can be consumed during the use phase of a building. Water use in buildings is typically categorized as blue, grey and black as shown in Figure 3.20. Significant efforts to reduce water use in

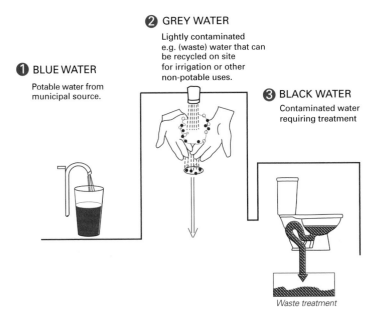

❷ GREY WATER

Lightly contaminated
e.g. (waste) water that can
be recycled on site
for irrigation or other
non-potable uses.

❶ BLUE WATER

Potable water from
municipal source.

❸ BLACK WATER

Contaminated water
requiring treatment

Waste treatment

3.20 Water use classification in building systems

buildings have focused on reducing the blue water consumed (low flow toilets and showers) and separating grey and black water systems (treating and re-using grey water on sites, reducing the need for blue water).

A comprehensive LCA would include the quantity of water consumed during energy production (e.g. in an industrial boiler). However, not all LCI databases include the water inventory for fuel combustion. Energy is consumed for water extraction, transformation meets water use standards, transportation and treatment. Desalination of water requires significant energy as well. LCA does not automatically track the energy impacts of water handling, as this would depend upon the goal and scope of the study. Better LCI data is required to integrate comprehensive water footprinting into practice.

In order to interpret the results of a water footprint, one needs to understand regional water quality and availability. Some studies (Hoekstra et al., 2011: 165) discourage the practice of applying weighting factors to water use at different regions because they feel it is subjective, not static (e.g. can depend upon time of year, etc.) and that weighted analysis can obscure the relevance of the total number, and they recommend assessing the local impacts as part of the qualitative analysis of a water footprint and provide some guidance on how to accomplish this.

Others are developing methods to assess the environmental impacts of water consumption. A method developed by Phister et al. (2009) evaluates damage to human health, ecosystem quality, and resources. In order to assess adequately, the type of use, source of water and geographic location needs to be reported along with the quantity of water consumed. Integrating water consumption and impacts in LCA is not yet standardized, and better data, tools and methods are needed to advance our ability to understand water-related impacts. At the present time, reporting consumption quantities is an appropriate first step to attain better data as the impact assessment methodology develops.

3.9 Waste generation

AS ASSESSING THE ENVIRONMENTAL IMPACT of waste treatment is difficult and uncertain, yet as the value of reducing waste is understood, the quantity of waste and materials for re-use and recycling is often reported as inventory items in LCA. These inventory items are reported in kg of material.

Calculating quantities is relatively straightforward, requiring an accurate counting of material quantities for disposal.

LCA typically classifies waste as hazardous, non-hazardous or radioactive. The definition of the waste types depends upon the region. In the US, the EPA is charged with developing requirements for management of hazardous waste and provides resources to evaluate waste. In order to establish if a material is hazardous, the EPA process consists of four questions (EPA, 2013c).

1   Is the material a solid waste?
2   Is it specifically excluded from regulation?
3   Is it a listed hazardous waste?
4   Does it exhibit a characteristic of hazardous waste?

Criteria 1–3 rely on specific EPA regulations that clarify known materials that meet the fourth criteria. The EPA defines the characteristics of hazardous waste, as shown in Figure 3.21.

IGNITABILITY – can create fires, is spontaneously combustible, has flash point less than 60 degrees Celsius.

CORROSIVITY – acids or bases capable of corroding metal containers.

REACTIVITY – unstable under "normal" conditions, can cause explosions, toxic fumes, gases or vapour when mixed with water.

TOXICITY – harmful or fatal when ingested or absorbed, can leak from waste to pollute groundwater.

3.21 EPA's hazardous waste characteristics
(EPA, 2013d)

Note there is not always consensus on whether or not a material should be considered hazardous waste. Fly ash is generated during the production of coal combustion for energy generation. Fly ash can be considered a waste product (as is typical in the US) or a by-product (as is typical in EU). Some are concerned that fly ash poses health risks and should be classified as a hazardous waste. The classification of materials results from a combination of science, policy and politics.

In Europe, the European Waste Framework Directive is developing lists of wastes and drafting legislation to codify European consensus on these issues (EUCE, 2013). This draft identifies 15 categories of hazardous waste:

1   EXPLOSIVE – explosions by chemical reaction that damage surroundings.
2   OXIDIZING – contributes to the combustion of other materials.
3   FLAMMABLE – readily combustible.
4   IRRITANT – can cause skin and eye damage.
5   ASPIRATION TOXICITY – causes toxic effect when inhaled to lungs via breath.
6   ACUTE TOXICITY – causes acute toxic effect if consumed or in contact with skin.
7   CARCINOGENIC – causes cancer or increases its incidence.
8   CORROSIVE – causes skin corrosions.
9   INFECTIOUS – contains micro-organisms or toxins which are believed to cause disease.
10 TOXIC FOR REPRODUCTION – has adverse effects on sexual function and fertility in adults and developing offspring.
11 MUTAGENIC – causes mutation that is a permanent change in the genetic material in a cell.
12 RELEASE OF TOXIC GAS – waste which releases acute toxic gases in contact with water or an acid.
13 SENSITIZING – wastes which contain one or more substances known to cause sensitizing effects to the skin or the respiratory organs.
14 ECO-TOXIC – presenting immediate or delayed risks for one or more sectors of the environment.
15 YIELDING ANOTHER SUBSTANCE – capable of exhibiting hazardous property listed above not in the original waste.

This European framework has not yet been adopted. Thus LCA practitioners must make their own judgement in assessing how to classify hazardous waste.

The International Atomic Energy Agency (IAEA) classifies radioactive waste, based on the level of hazard and length of time to decay. LCA reports all radioactive waste as a single inventory item and should include all wastes that contain radioactive material. The primary source of radioactive waste from buildings and building products is as a result of nuclear power generation.

Naturally occurring radioactive materials are concentrated in the following industries (WN, 2013):

- the coal industry (mining and combustion);
- the oil and gas industry (production);
- metal mining and smelting;
- mineral sands (rare earth minerals, titanium and zirconium);
- fertilizer (phosphate) industry;
- building industry (granite and other materials);
- recycling.

Buildings can also contain radiation based on exposure of naturally occurring radon in the ground.

3.10 Material re-use and recycling

MATERIAL AVAILABLE FOR RE-USE AND RECYCLING (kg of total material) is an inventory item that can be reported in LCA. These quantities attempt to capture the efficiency of the manufacturing or complete life cycle of products and can be compared with the quantity of waste materials generated.

Recycling and material re-use can be consumed within the process being studied (e.g. scrap metal returning as feedstock). Typically this material re-use results in a reduced requirement for input material and thus reduced LCA results. See Section 6.2 for more detail on the methodological challenges of integrating recycling impacts into LCA.

During the full life of a building or product, materials may be available for re-use (crushed concrete as a roadway bed) or recycling (scrap steel used in rebar manufacturing). In the former, the material has lower performance characteristics (down-cycled). In the later, recycled steel can be used to create as strong or stronger structural steel shapes (up-cycled).

Reporting the material available for re-use and recycling does not quantify environmental benefits directly but can help assess total impacts. When including use stage and end-of-life stage impacts, the percentage of material re-used or recycled should reflect actual in practice quantities rather than the maximum possible value. For example, just because reinforcing steel can be recycled, does not mean it will all be. Much of it remains bonded with concrete and left in place.

chapter 4

Life Cycle Assessment Data

DATA AVAILABILITY AND MANAGEMENT are critical challenges in the development of environmental life cycle assessments that are efficient to perform and contain meaningful results. The inventory analysis is the primary data collection stage of an LCA. As described in Section 2.3, this stage requires the assessment of quantities of material and energy used and access to LCI databases to determine the appropriate emissions for each of these activities.

Determining the quantity of materials and energy used requires access to manufacturing and construction data or the expertise to estimate them. LCI data is available in a variety of formats from both open access and proprietary sources. Unfortunately, current databases are incomplete and are not detailed enough to capture many project-specific conditions. A simple manufacturing process may still include dozens of manufacturing steps and track hundreds of different emissions. Organizing and compiling the data require a rigorous methodology and attention to detail.

Building industry professionals who use LCA in their practice are likely to do so in one of the following frameworks:

- interpreting LCA results prepared by others;
- commissioning an LCA to be prepared by LCA expert;
- using a building-specific LCA tool to study design options;
- building a simple LCA model for analysis.

This chapter provides an outline of data sources and calculation methods appropriate for building professionals looking to make an informed interpretation of LCAs and apply LCA within design and construction practice.

THE DETAIL AND PRECISION NEEDED to develop the material and energy inventory are directly related to the goal and scope of the study. Defining the system boundary appropriately requires an understanding of the limitations of available time and data as well as the relative importance of different processes. More care should be taken in getting high quality data for the processes with the largest contribution to the total environmental impacts. If the impacts of the process are "not significant", the process can be omitted from the study (see Section 4.5).

LCA "cut-off rules" define the limits of significance (e.g. processes contributing less than 1 per cent of the energy or mass of a product can be considered insignificant) for an LCA and enable items to be omitted from the study. The concept of cut-off rules creates a perplexing paradox: one can omit analysis of items with small impact, but one has to analyze the items first to understand their significance.

How does one know whether or not a process is significant without first doing the LCA analysis?

One method of understanding significance is to recognize that LCA is often performed in an iterative manner, ranging from quick and general to detailed and comprehensive. Figure 4.1 represents the relationship between

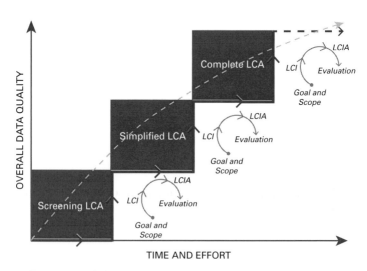

4.1 Iterative nature of LCA

(Adapted from ILCD, 2010a: 25)

time and effort and overall data quality as described in more detail in the *ILCD Handbook* (ILCD, 2010a), which was developed by the European Commission's Institute for Environment and Sustainability to provide guidance for LCA practitioners.

Building industry professionals should recognize that there are usually significant limitations to quick LCA studies and that the detail and precision may not be adequate to make definitive decisions. Terms outlined in more detail in the European EeBGuide Project (Wittstock *et al.*, 2012), to characterize the comprehensiveness of an LCA study (screening, simplified or complete) have been adapted here to help characterize different levels of refinement. The EeBGuide program entitled "Operational Guidance for Life Cycle Assessment Studies of the Energy Efficient Buildings Initiative" has developed specific recommendations for each of these LCA levels for whole building and product LCAs.

Screening LCA

A screening LCA is performed to develop estimates of the environmental performance of a system. It provides a quick, and economical understanding of a product or a building. This level of analysis, though not generally detailed enough to make definitive recommendations, can be used to identify "hot spots" (the materials, processes or life cycle stages with significant environmental impact and which are worthy of detailed analysis).

Industry-specific LCA tools that use predefined LCI datasets interfacing with user input of material and energy quantities provide an efficient method for designers to integrate screening LCAs into their design process. Economic Input–Output (EIO) analysis (see Section 4.3) also provides an effective mechanism to perform a high-level screening LCA for complex systems.

Simplified LCA

A simplified LCA could be the stepping-stone to a complete LCA or a focused study on one or more aspects of the system under study. In some instances, as quantities of materials are not known, the data is omitted. Simplified LCAs might exclude the contributions of a known process (e.g. mechanical equipment manufacturing in a whole building LCA) based on the results of previous studies but without including the technical justification that the impacts of the mechanical equipment are not significant and that the LCA cut-off rules are satisfied.

Most LCA studies, even those that are quite detailed, performed by building industry professionals would be considered screening or simplified LCAs. Developed without access to sophisticated LCA tools and with less understanding of LCA standards and practices, these studies are most appropriate for use in advancing internal understandings of environmental impacts and exploring methods to improve practice. They can be the foundation for a more detailed and complete ISO-compliant LCA.

Complete LCA

A complete LCA is often the result of an iterative LCA. Refining screening and simplified LCAs to include complete inventories and high quality data enables a more accurate and comprehensive analysis. An ISO-compliant LCA should be complete, and provided it covers the entire product life cycle, could enable comparative assertions or public publication (see Section 5.1).

4.1 Inventory input: material and energy

IN ORDER TO QUANTIFY the environmental impacts of processes, the quantities of materials and energy consumed during the life of the product system must be determined. The data can be primary (collected for the process under study) or secondary (collected by someone else and used for the analysis). Measuring, modeling, researching and estimating are all methods to determine the input quantities for an LCA.

- Measuring data: Material and energy use can be quantified by reviewing purchasing records (e.g. kg of steel purchased), and utility bills (e.g. MJ electricity purchased) or by physically measuring quantities (e.g. m^3 of water pumped from the ground).
- Modeled data: An LCA performed during the design stage of a project would not have documentation of actual quantities that will be used. Analytic models (e.g. energy performance models) can be used to predict consumption. Manufacturing systems can also be modeled based on engineering principles (e.g. understanding a specific chemical reaction, the energy and material input can be modeled per unit of output desired).
- Researching published data: Government and industry surveys can provide detail on typical consumption practices (e.g. typical transportation

distances and modes) and regional variation of electrical energy sources. Future life cycle stages can be estimated using historical data and surveys (e.g. recycling rates for a specific material).

■ Estimated data: In some instances the quantities must be estimated using judgement and experience. The LCA could include estimates of material quantities used (e.g. a drywall subcontractor could provide guidance on the extra quantities of wallboard typically purchased to account for waste and fit-up) or estimates use scenarios (e.g. development of maintenance and repair schedules for an office building).

CASE STUDY 4.1: WHOLE BUILDING MATERIAL QUANTITIES

In order to conduct a whole building LCA including the impacts related to material manufacturing and construction, the quantities and types of material used must be determined. Depending on the stage of project, the access to quantity data and expertise of the LCA practitioner, different methods can be appropriate to use to attain this data.

■ Quantity take-off: Similar to the work done by a quantity surveyor or construction cost estimator, the materials for construction can be defined by creating a detailed list of materials and their quantities, the quantity take-off (QTO). Fairly complete construction documents are required to perform this task unless the estimator has the ability to predict material needs and specifications. Transportation distances and modes can be based on knowledge of project specifics or using industry average data. Tabulating material quantities can be a time-consuming effort and is most efficient if it can be integrated into the existing cost estimating methodology or can take advantage of data included in the construction document building information model (BIM), as noted in Section 7.2.

■ Subcontractor surveys: During or after construction, the quantity of materials used on the project can be estimated through surveys of the primary subcontractors. This method has the advantage of representing the total materials purchased, not just those installed, and thus accounts for standard practices of material waste. Additionally, the surveys can determine the actual distance and modes of travel used to bring the goods to the job site.

■ LCA tools for buildings: LCA tools have been and continue to be developed to help building professionals develop LCAs of whole buildings in a timely and cost efficient manner. These tools can bridge between general building descriptions (e.g. a concrete floor of a certain size and span) to estimate material quantities (e.g. kg of concrete, reinforcing steel, and floor finishes) based on industry-standard construction practices. Other tools are developing that extrapolate the quantity data from BIM models and translate to inventories relevant for LCA (See Section 7.2).

■ Estimate from professional judgement: An experienced professional (or team of experts) can estimate material quantities of projects even only schematically defined. Structural engineers have knowledge of typical weights of steel per unit area for typical construction types. Architects could estimate the percentage of glazing and typical wall assemblies when detailed elevations are still in development. Quantity surveyors and construction cost estimators with experience developing costs estimates during the design process use similar skills to estimate costs before projects are fully defined.

■ Construction costs: As noted in the QTO section, the construction cost estimate often includes significant information about the materials and their quantities. In larger construction projects, the general contractor often builds up the total construction cost from subcontractor pricing for the majority of the construction items and thus the cost schedule does not include detailed breakdowns of material quantities. However, the EIO method of LCA (see Section 4.3) provides methods to correlate dollars spent in an industrial sector to environmental emissions, and a detailed construction cost report can function as the primary material inventory for an LCA.

4.2 Inventory emissions: unit process data

EMISSIONS INVENTORIES ARE THE COMPILATION of LCI analysis results and report the consumption and emissions resulting from a specific process. Typically an LCI is created based upon an LCA that was designed to create the inventory and is produced based on as much primary data as is practical. LCI databases are compilations of multiple studies and can be public and free to access or private non-profit or proprietary and included with LCA software. There are two types of LCI databases: (1) process LCA databases (a collection of the results of past studies and the primary source of emissions data for processed based LCA); (2) EIO LCA databases (linking environmental impacts to economic sectors as discussed in Section 4.3).

Note data collection and reporting from industry are voluntary and thus LCI databases are not necessarily comprehensive or up to date. Private databases tend to be more complete as access fees can be used to support locating, updating and refining the LCI datasets. Government support can be key to developing good LCI data. The US Department of Agriculture has funded LCI studies resulting in more developed databases for agricultural products (including bio-based building products) than exist for other building products. Additional government funding to support the development of high quality LCI data for a greater range of building materials, processes and products would help improve the quality of LCA analysis of buildings.

LCIs can be published for a single process or aggregated for a product. Different industries publish data in differing levels of aggregation. Publishing the individual process data permits LCA professionals to customize LCA studies. Publishing aggregated data makes it easier for non-LCA experts to use and interpret and ensures that the LCI is modeled to reflect industry practices for a specific material or product. Currently, the US LCI has over 175 unit process inventories related to wood product manufacturing and there are ten inventories for steel processes. World Steel publishes aggregated data for individual types of steel products as the industry believes this is the most accurate and useful. LCI data for wood harvesting and product manufacturing is published as both individual unit processes as well as aggregated LCI results for products such as a glue laminated beam.

CASE STUDY 4.2: LCI OF KILN DRYING PROCESS

Table 4.1 shows a summary of the inventory for the single process of kiln drying lumber. The input of 435.78 kg of rough sawn lumber combined with diesel, oil, combusted wood and purchased electricity results in four primary emissions as well as the product of 1m^3 of kiln dried lumber. Note that the emissions related to producing inputs, including the rough sawn lumber as well as combustion of fuels to make the energy, are not included in this unit process inventory. In order to understand the total emissions cradle-to-gate of kiln dried lumber this LCI would need to be combined with those for the input materials and energy.

Figure 4.2 is a simplified diagram showing the different unit processes included in the LCA stage of glue laminated beam manufacturing. This data could be presented as a single aggregated data set or as nine different unit process data sets.

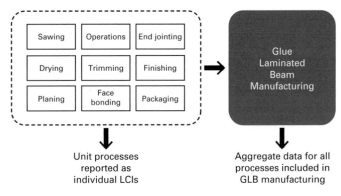

Unit processes reported as individual LCIs

Aggregate data for all processes included in GLB manufacturing

4.2 Unit process compared to aggregated LCI data

Table 4.1 **LCI for kiln dried lumber**

Inventory Name	Quantity
Sawn lumber, softwood, rough, green, at sawmill, INW	435.78 kg consumed
Diesel, combusted in industrial equipment	0.13 l consumed
Residual fuel oil, at refinery	0.0776 l consumed
Wood, softwood, INW, generated at lumber mill, combusted in industrial boiler	55.3 kg consumed
Electricity, at grid, Western US, 2000	16.7 kWh consumed
Natural gas, combusted in industrial boiler	25.6m³ consumed
VOC, volatile organic compounds	0.168 kg emitted
Formaldehyde	0.00064 kg emitted
Acetaldehyde	0.00201 kg emitted
Methanol	0.0225 kg emitted
Sawn lumber, soft wood, rough, kiln dried at kiln, INW	1m³ produced

(NREL, 2013)

4.3 Inventory emissions: economic input–output data

ECONOMIC INPUT–OUTPUT LCA (EIO-LCA) is a different method of establishing emissions and analyzing LCA data. The LCA inventory developed, as noted in Sections 4.1 and 4.2, based on developing data for individual manufacturing processes is called the process-based LCA method (often shortened to LCA).

EIO-LCA uses economic data as the foundation to establish life cycle inventories. The initial economic model was proposed in the 1930s by an economist who proposed compilation sales from one sector to another (Leontief, 1986) into a matrix model. Today the economic flow data produced by the US Bureau of Economic Analysis (BEA) has been combined with industrial emissions data to create EIO-LCA databases. Carnegie Mellon University's Green Design Institute (CMU GDI) created the first such database in the mid-1990s.

The EIO analysis divides the economy into different sectors such as "wood windows, doors and millwork", "flat glass manufacturing", or "dog and cat food manufacturing" that tracks sales between sectors. For example, in order to purchase one million dollars worth of wood windows, a certain quantity of flat glass must be purchased. An EIO estimates the production required in all the sectors of the economy to support increases in output of any given sector. An EIO-LCI adds environmental emissions for the activity in each sector, resulting in the ability to compute emissions from an increase in spending/production within one sector that includes all the upstream production impacts (Hendrickson et al., 2006).

The US government currently categorizes the US economy into nearly 500 industrial sectors. Other governments typically publish less detailed data. EIO-LCA databases developed by academia (CMU GDI, 2008) and private industry (CEDA, 2011) enable an LCA to be performed based on dollars spent in each sector without requiring the detailed quantities of materials and energy consumed, as required in a process-based LCA.

CASE STUDY 4.3: EIO-LCA DATA

The comprehensive nature of EIO-LCA can be seen if reviewing the LCA of flat glass manufacturing sector. The Green Design Institute at Carnegie Mellon provides an open access EIO-LCA tool (www.eiolca.net) .providing data regarding economic activity, conventional air pollutants, GHGs, energy, hazardous waste, toxic releases and water withdrawals.

Using the EIO-LCA.net online tool, the GHG impacts due to one million dollars of spending in flat glass manufacturing (CMU GDI, 2008) were determined and are summarized in Table 4.2. Note that the largest contribution is in the flat glass manufacturing itself. This represents the fact that glass manufacturing consumes a significant amount of energy and manufacturing facilities are combusting a significant amount of fossil fuels on site. Off-site power consumption is the second largest contributor to GHG emissions. Note the emissions from "oil and gas extraction" have significant contributions from process emissions (chemical reactions during the manufacturing process) and the release of other GHGs (dominated by methane that is released during the production of natural gas).

Other sectors such as "sand mining" and "transportation" are directly attributable to the manufacturing of glass. However, in order to make glass, one needs manufacturing equipment and facilities which likely resulted in "iron and steel mills" and "cement manufacturing" sectors having a significant impact. These 20 sectors contribute to 95 per cent of the total GHG emissions. Expanding the search to include all sectors results in increasingly smaller contributions from a wide range of sectors including "car washes" and "electric lamp bulb and part manufacturing" (a glass factory would need light bulbs and would wash their equipment). Even these smaller inputs can be captured by EIO-LCA. They are most commonly excluded from the system boundary in process-based LCA, given the impracticality of collecting the data for analysis.

EIO-LCA databases and tools can report emissions per sector, creating an LCI for each sector, or internalize the impact characterization (see Section 4.4) to report environmental impacts.

An EIO-LCA is particularly suited to performing a screening LCA of complex systems such as a large corporation or building. Leveraging existing high-level financial data such as annual reports or construction cost estimate, an LCA can be performed by dividing the spending into appropriate industrial sectors without requiring a detailed understanding of the exact material quantities and processes that take place.

Table 4.2 Top 20 sectors GHG emissions for one million dollars of flat glass production

Sector	$(TONS\ CO_2e)$			
	TOTAL	Fossil	Process	Other
Flat glass manufacturing	1090	1090	0	0
Power generation + supply	489.0	481.0	0	7.42
Oil and gas extraction	71.5	20.1	13.1	38.3
Coal mining	52.5	5.93	0	46.6
Alkalines + chlorine manufacutrin	48.0	48.0	0	0
Pipeline transportation	29.5	13.5	0.037	16.0
Truck transportation	27.5	27.5	0	0
Rail transportation	26.8	26.8	0	0
Petroleum refineries	21.8	21.8	0	0.068
Natural gas distribution	21.5	1.94	0	19.5
Other organic chemical manufacturing	16.4	14.8	0	1.69
Sand, gravel, clay + refractory mining	15.1	15.1	0	0
Industrial gas manufacturing	14.7	1.70	0	13.0
Iron + steel mills	12.1	4.57	7.46	0.074
Clay + non-clay refractory manufacturing	9.69	9.69	0	0
Waste management + remediation services	8.15	0.298	0	7.86
Fertilizer manufacturing	6.63	1.64	2.22	2.77
All other basic inorganic chemical man.	6.37	3.85	2.52	0
Other non-metallic mineral mining	5.94	5.94	0	0
Cement manufacturing	4.96	2.07	2.89	0
TOTAL (tCO_2e):	2050	1850	31.6	159.4

(CMU GDI, 2008)

CASE STUDY 4.4: EIO-LCA HYPOTHETICAL BUILDING

The LCA calculations shown in Tables 4.3 and 4.4 were performed in approximately two hours using the given cost estimate and the academic version of the CEDA EIO-LCA database (CEDA, 2011). The building is a hypothetical building and costs used are for demonstration only. The emission factors are placeholder values representing the relative range of carbon dioxide equivalents for different industries (as the CEDA database is proprietary). They are presented for demonstration purposes only and should not be used for other analyses.

Table 4.3 EIO-LCA calculation summary

Sector	Total Cost	Percentage Labor	Material Value (2002 dollars)
Overhead, insurance	$1,800,000	99%	$11,664
General conditions	$2,400,000	80%	$311,040
Earthwork	$1,500,000	60%	$291,600
Concrete	$3,000,000	40%	
6%			$69,984
47%			$676,512
47%			$443,232
Wood + plastics	$1,200,000	40%	
60%			$279,936
40%			$186,624
Plumbing	$3,600,000	40%	
15%			$209,952
5%			$69,984
20%			$279,936
40%			$559,872
20%			$279,936
Thermal protection	$1,800,000	40%	$699,840
Openings	$4,800,000	40%	
50%			$933,120
50%			$933,120
Metals	$3,600,000	40%	$1,399,680
Finishes	$1,500,000	40%	
60%			$349,920
30%			$174,960
10%			$58,320
Specialties	$300,000	60%	$77,760
Conveyor	$300,000	60%	$77,760
Fire suppression	$300,000	50%	$77,760
Electrical	$2,700,000	50%	
40%			$349,920
60%			$524,880
Utilities	$900,000	40%	$349,920
Site improvements	$300,000	80%	$38,880
TOTAL:	$30,000,000		$9,789,984

Table 4.4 EIO-LCA calculation summary

Industry Sector	Approx. CO_{2e} emissions factor	Subtotal	Total kg
Management of companies + enterprises	0.1		**1166**
Management of companies + enterprises	0.1		**31,104**
Truck transportation	0.5		**145,800**
Concrete			**3,837,456**
Veneer and plywood manufacturing	0.5	34,992	
Ready-mix concrete manufacturing	3	2,029,536	
Iron + steel mills + ferroalloy manufacturing	4	1,772,928	
Wood + plastics			**513,216**
All other misc wood product manufacturing	0.5	139,968	
Plastics material + resin manufacturing	2	373,248	
Plumbing			**1,084,752**
Plumbing fixture fitting + trim manufacturing	0.5	104,976	
Fabricated pipe + pipe fitting manufacturing	1	69,984	
Custom roll forming	1.5	419,904	
Air conditioning, refrigeration, + heating equip, manufacturing	0.5	279,936	
Heating equipment (no furnaces) mfrg	0.75	209,952	
Polystyrene foam product manufacturing	1	699,840	
Thermal protection			**3,265,920**
Aluminium product manufacturing	1.5	1,399,680	
Flat glass manufacturing	2	1,866,240	
Iron + steel mills + ferroalloy manufacturing	4	5,598,720	
Openings			**2,420,280**
Lime + gypsum product manufacturing	7	2,099,520	
Brick, tile + other structural clay product manufacturing	1.5	262,440	
Paint + coating manufacturing	1	58,320	
All other miscellaneous manufacturing	0.5	38,880	
Motor + generator manufacturing	0.5	38,880	
Other electronic component manufacturing	0.5	38,880	
Electrical			**437,400**
Other electronic component manufacturing	0.5	174,960	
Lighting fixture manufacturing	0.5	262,440	
All other miscellaneous manufacturing	0.5	174,960	
Truck transportation	0.5	19,440	
TOTAL:			**11,737,094**

The steps taken to develop this EIO-LCA are summarized as follows:

1   Obtain cost estimate by specification section.
2   Estimate the percentage of construction costs related to materials, as EIO is material-based and LCA typically does not attribute environmental impact to labor.
3   Reduce costs by estimated profit.
4   Subdivide sections with multiple materials into approximate spending on different materials.
5   Deflate values to convert to dollar value to match the year of a given EIO database.
6   Identify the most appropriate industrial sector for the tracked material expenditure.
7   Obtain emissions factor (estimate of emissions per dollar spent in sector) from the EIO-LCA database.
8   Determine total kg CO_2e by multiplying 2002 spending by emission factors.

The accuracy of this analysis depends on key decisions related to estimating labor percentages and selecting the appropriate industrial sectors. Interesting choices included how to account for earthwork, which will include significant use of heavy equipment. In this example, we selected truck transportation for a proxy value. How should the environmental impact of overhead and general conditions be modeled? In this example, we selected the sector "management of companies and industries". Note that the emissions factor is relatively low for company management. This screening level LCA was developed relatively quickly and can help to target where more detailed study is warranted.

Developing a process-based LCA would require significantly more time to estimate material quantities and link to environmental impacts unless a simplified building industry tool is used to estimate material quantities based on building geometry.

Often items (such as the MEP systems) are omitted from conventional process-based LCA due to the difficulty in estimating the material quantities. This case study estimates that the MEP systems contribute less than 10 per cent of the total embodied impacts, which is significant if not substantial.

At many stages of the design and construction process, costs are known when quantities are not. In these cases, the EIO-LCA method has several advantages. Based on this quick analysis, the total embodied carbon footprint (and

similarly other LCA impacts with slightly more effort) as well as relative contributions to the embodied carbon footprint can be estimated. Figure 4.3 shows
the breakdown of embodied carbon footprint per section of the construction cost
estimate identifying that concrete, metals and openings are the three largest
contributors.

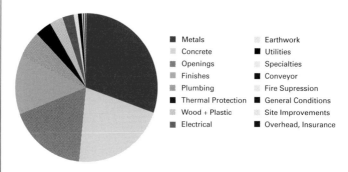

4.3 EIO-LCA results from hypothetical building

Comparison of process-based and EIO-LCA

Process-based LCA is the traditional method for inventory assessment. As
described in previous sections, process-based LCA consists of methodically
analyzing material and energy flows at every stage of the life cycle to understand precise consumption and emission values. This method in practice is
time-consuming, often limited in scope to a handful of products or processes,
and difficult to scale to complex systems.

In contrast to process LCA, EIO-LCA starts with a model of the
national economy. EIO tables and industry-level environmental data are used
to construct a database of environmental impact per dollar of sales from an
industry. This method is considered more comprehensive than process LCA
because the EIO tables capture the interrelations of all economic sectors;
however, input–output LCA has the problem of providing only aggregate
industry-level data and is based upon 2002 economic data.

Advantages of EIO-LCA

- Fast screening level LCA of complex systems.
- Ability to leverage existing economic data (e.g. annual reports and cost estimates).
- Includes comprehensive supply chain inputs beyond what is practical for most process based LCAs.
- Can provide strong foundation for hybrid LCA (see next section).

Disadvantages of EIO-LCA

- Unable to conduct comparative analysis within a sector as data is aggregated by industry.
- 2002 economic data is definitely significantly outdated for sectors (e.g. technology) that have completely changed since that time.
- Imported products typically are modeled as like products manufactured in the US.
- Specificity limited by the limited granularity of the EIO tables.

Given that both process-based and EIO-LCA methods have unique strengths and weaknesses, a hybrid approach can be used to combine the two methods to minimize the weaknesses of each and take advantage of their strengths.

Hybrid LCA

Hybrid LCA is a structured approach to combining process-based and EIO-LCI data. There are two basic types of hybrid methodology that differ in the order of collecting the data: (1) start with primary and process LCI data and fill in data gaps using EIO-LCA data; or (2) start with EIO-LCA data to evaluate hot spots and then collect primary and LCI process data to refine the results where necessary. The first approach is often called "bottom-up" LCA and the second approach "top-down" LCA.

The risk with bottom-up LCA is that the practitioner may spend time collecting primary data for components of the analysis that turn out to be insignificant. The risk with top-down LCA is that the practitioner could reach a false conclusion based on misapplication of data. In general, process LCA is best applied to relatively contained studies, and hybrid analysis to larger, more complex studies.

4.4 Analysis: attributional vs. consequential

IN DEVELOPING AN LCA MODEL, the methodology must be consistent. There are two distinctly different methods to model life cycle inventories in LCA: attributional and consequential. The attributional method is the most common and simpler method to use. The following is a summary of these different methods. More detail is presented in the *ILCD Handbook* (2010a: 70–72).

Attributional LCA uses current and historical data that is measurable/known to model the impacts of processes throughout a product's life cycle. The model assumes that changes within the LCA system do not impact the overall technosphere (the interaction between technology and the environment). For example, an attributional LCA model comparing wood and steel structures presumes the current manufacturing methods, costs and supply chain for both materials.

Consequential LCA attempts to model the future conditions, that might be expected to adapt as a consequence of changes to market conditions, modeling the consequences of different actions. A consequential LCA comparing the impact of building more commercial buildings out of wood would need to look at the consequences of increased market demand for wood products.

Attributional LCA is most appropriate when studying a confined system that is small relative to the total economic and environmental system: when changes to the system under study will not likely have any larger impact. Consequential LCA is appropriate when the LCA study is used to make large-scale policy decisions or major shifts of manufacturing process that will have a consequence on the larger economic and environmental systems. For example, an attributional LCA comparing the benefits of bio-fuels might be appropriate to influence decisions of a small manufacturer. However, if that study is scaled up to national policy, the consequences of substantially increasing demand for bio-fuels (cutting down forests, converting farmland from food to bio-fuels) would need to be evaluated. Consequential LCA enables the integration of future projections into the LCA analysis.

4.5 Analysis: data quality

DATA QUALITY MUST BE ASSESSED in order to interpret LCA results correctly. As is noted in other sections, the development of an LCA often requires the assembly of data from diverse sources to model the system under study. There are both qualitative and quantitative measures of LCI data quality. ISO 14044 requires that the data be evaluated on multiple criteria (ISO, 2006b) that are categorized into simplified groupings:

- representativeness of time, geography and technology;
- data use: precision, completeness and consistency;
- reproducibility and data sources;
- uncertainty.

Representativeness is the evaluation of how representative the LCI data is with regard to technology, geography and time. The technological representativeness should address the process (e.g. hot rolled vs. cold rolled steel) and the material and fuel source (e.g. coal vs. natural gas). Geographic representativeness reflects how accurately the data used reflects the actual conditions (e.g. LCI data for global steel production vs. specific data for the region under study) or the location under study (e.g. chemical admixture data from Europe used in an LCA of concrete produced in the US). As processes and efficiencies are continuously developing, the time of the data collection impacts the representativeness of the data used (e.g. data from 20 years ago may not be representative of current conditions). Especially when predicting consequences and future conditions, care must be taken to evaluate the current and potential changes to emissions when extrapolating from current or past data.

The use of data should be assessed with regard to its precision, completeness and consistency. Precision is the evaluation of how variable the data is. Typically LCA data represents average data. Precise, average data would mean that there is little variation between the averaged data sources and thus the average data very likely represents the actual data. Imprecise data would have a large variation and thus average data may not be representative of actual conditions.

CASE STUDY 4.5: IMPRECISE AVERAGE DATA

In many conditions, industry average data is used for LCA analysis. Figure 4.4 represents hypothetical energy use from an industry survey of 50 manufacturing facilities. In this industry there are two primary production methods to produce the same product. Process A is significantly more efficient than Process B.

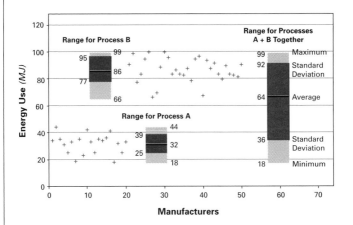

4.4 Energy use of 50 hypothetical manufacturers

In this example, the industry average (typically weighted in proportion to production volume) is 64MJ, with a standard deviation of 28. If this average was used to estimate emissions for a specific product, the data would not represent actual technology and would be imprecise: there is a large variability in the actual data. More precise data would represent the actual technology used, selecting average data from either Process A or Process B, increasing precision.

Completeness assessment evaluates if the analysis and data are adequate to support the goal and scope of the study and ISO requires a justification of any omissions (ISO 2006b: 26). Recognizing that obtaining comprehensive and complete data for an LCA is challenging, the standards permit the reporting of incomplete data but require that the goal and scope be evaluated, and perhaps adjusted, to reflect the reality of the data available. And critically, the evaluation and any justification for omissions must be clearly documented and included in the LCA report. Assessing the relevance of unknowns is quite challenging and requires professional judgement and expertise.

Consistency checks, evaluating both qualitative and quantitative aspects of the data, are an evaluation of the assumptions, data and methods to ensure they are consistent with the established goal and scope. This check should evaluate data quality throughout the analysis to see if inconsistencies might result in inappropriate conclusions. And critically, allocation methods must be consistently applied as outlined in Section 4.6 and Chapter 6.

An LCA should be reproducible (an independent analysis based on the same assumptions, should turn in the same results) based upon the information provided in the LCA report. Therefore, information must be provided about how data was collected or estimated and sources of the data included within the documentation.

Uncertainty, though important to understand when interpreting LCA results, is difficult to characterize. As outlined in the *ILCD Handbook: General Guide for Life Cycle Assessment* (ILCD, 2010a: 377–380), uncertainty can be classified as random (related to random system behaviour), choice (related to decisions made when conducting the LCA), and ignorance (related to lack of knowledge about the studied system). These uncertainties can manifest in:

- the inventory analysis: quantifying material and energy use;
- the LCI data: resource use and emissions for each process;
- the LCA methods: assumptions when establishing the study.

See Chapter 6 for examples of uncertainty in building industry LCA and the implications of uncertainty on the ability to act upon the results of an LCA. Some LCA tools include statistical methods to calculate uncertainty over multiple processes using simulation models such as "Monte Carlo" analysis to estimate the aggregated uncertainty of a system under study. Understanding uncertainty, even just the range between average, maximum probable and minimum probable results, is quite useful in evaluating an LCA. Unfortunately, most building industry LCAs report industry average data and rarely include statistical ranges of uncertainty or variability.

As outlined in Section 2.5, interpreting LCA results requires an evaluation to understand if the results are sensitive to variation in a specific data set. If the data quality for a specific process is poor, but the impact of that process is low, getting higher quality data may be a low priority. When refining

4.5 Data quality and sensitivity decision matrix
(Adapted from ILCD, 2010a: 298)

LCA data, increasing the quality data for high significance processes should be prioritized, as shown in Figure 4.5.

4.6 Analysis: allocation

MANY MANUFACTURING PROCESSES produce more than one product. Decisions must be made in setting up the LCA modeling methods to determine of the total emissions emitted in producing both products, what total emissions and wastes are allocated to each of the products. Figure 4.6 illustrates two products produced at one manufacturing facility with multiple inputs from and outputs to nature. These two products are considered co-products.

In some cases a factory has two distinct production lines. While production is in a shared facility, the production of one product has no influence on the production of the second product. An example of this is a tool manufacturer with two products, hammers and wrenches, produced with different equipment on different production lines as represented in Figure 4.7.

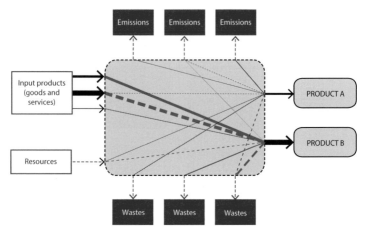

4.6 System boundary of multiple product system

(Adapted from ILCD, 2010a: 80)

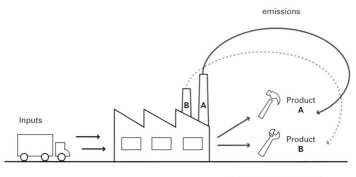

ALLOCATION STRATEGY:
divide unit process

4.7 Two independent co-products

In other processes, one product's production is the result of the manufacture of a primary product and it is called a by-product. By-products typically are secondary to the primary product in financial significance to the company and cannot be produced independently of the primary product. Asphalt is a by-product of refining petroleum. A by-product can be considered useful or as waste. ISO standards do not make distinctions between these different conditions, defining both as co-products.

An interesting LCA challenge occurs when uses are found for materials that had been or are currently considered waste. The sawing and planing of lumber result in wood chips. Instead of disposing of this "waste", the chips are used as fuel, either on site or sold. Thus this "waste" becomes a co-product and sometimes is identified as a secondary fuel. See Chapter 6 for more information on the challenge of determining if a product is a waste or co-product for LCA analysis.

Establishing how to allocate environmental impacts from manufacturing processes that result in more than one product can be quite challenging. ISO 14044 provides guidance on allocation methods (ISO 14044.4.3.4.2, 2006b) recommending a "stepwise" procedure as illustrated in Figure 4.8.

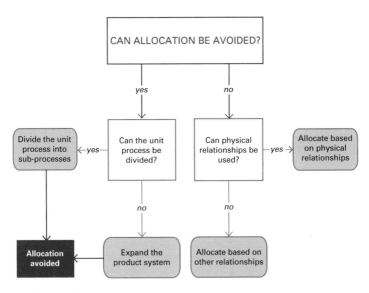

4.8 Allocation decision tree

CASE STUDY 4.6: ALLOCATION OPTIONS PER ISO

As noted earlier, ISO outlines a decision procedure for determining how to allocate emissions and impacts in LCA (ISO 2006b: 14). This section provides an overview of the process and case studies to help illustrate the process shown in Figure 4.8.

STEP 1: WHEREVER POSSIBLE, AVOID ALLOCATION

ISO outlines two primary methods to avoid allocation, either dividing unit processes or system expansion (ISO 2006b: 14), thus expanding the system boundary to include the additional functions of the co-products, or products produced together with another product.

Unit processes can be divided by collecting the input and output data for two products separately using methods such as sub-metering electricity supplied to manufacturing equipment or collecting input and output data separately. As represented in Figure 4.9, two separate LCAs can be performed as the input and output quantities of each product are independent.

When system expansion is used, the LCA scope of the first product is "expanded" to include the function and typical emissions related to the second product. This can either mean that the impacts "avoided" by producing the

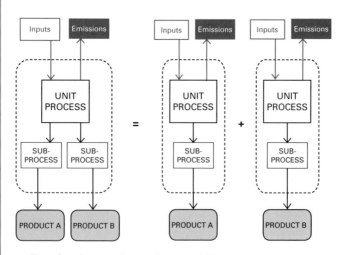

4.9 Two independent co-products resulting in two LCAs

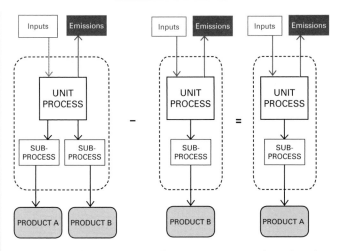

4.10 Subtracting the LCA of a co-product to estimate impacts of a single product (system expansion)

"co-product" are subtracted from the total amount or the functions of the co-product are added to the LCA data (see Figure 4.10).

For the example of steel manufacturing described in more detail in Section 6.1, the system could expand to include both the function of producing structural steel and cementitious materials made from the slag. The emissions avoided due to the reduction of cement produced could then be subtracted from the total emissions related to steel (and the co-product slag) production.

This method works well for cases such as glue laminated beam manufacturing where the co-product of wood chips is used as a fuel source to directly replace purchased fuels in the next stage of the manufacturing process. This method is less clear when applied to situations such as the slag described above where slag reduces but does not typically replace cement.

One difficulty with this method is that it presumes that the production of the co-product will ensure that the emissions related to its substituted product will not happen. This implies a consequence (change in future event) that is not certain. Depending on the application, this substitution may be more or less credible.

STEP 2: ALLOCATED BASED ON PHYSICAL RELATIONSHIPS

If allocation cannot be avoided, ISO recommends (ISO 2006b: 14) that LCA data should be allocated based on physical relationships if possible. Examples of physical relationships include: by mass (e.g. weight of product produced in a steel forge or per volume of product shipped by a truck when the quantity of material shipped is governed by the size of the material rather than the weight) (see Figure 4.11). An example of this allocation method is as used for the LCI of glue laminated timber discussed in earlier chapters. In this case the impacts related to growth and logging of trees are allocated in proportion to the mass between the multiple products of rough sawn lumber as well as the wood chips and sawdust (used as bio-fuels).

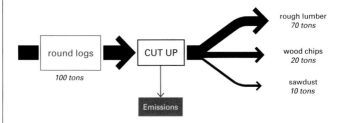

$$EMISSIONS_{product} = \frac{weight_{product}}{total\ weight}\left(\begin{array}{c}total \\ emissions\end{array}\right)$$

$$EMISSIONS_{lumber} = \frac{70}{100}\left(\begin{array}{c}total \\ emissions\end{array}\right)$$

4.11 Allocation by mass

STEP 3: ALLOCATED BASED ON OTHER RELATIONSHIPS

Allocation by other relationships occurs when it is not possible to allocate by any of the previously described methods (ISO 2006b: 14). The most common "other relationship" is economic value. In this case, the emissions are distributed in proportion to the value of the different products. A prime example of a system for which financial value is an appropriate allocation method is coal power generation and the by-product of fly ash (a combustion residual that can be a substitute for cement). The units of product produced are not compatible: energy is produced in kJ and fly ash is produced in kg. The production of energy is the primary function of the plant. The emissions related to production could be allocated between the different processes in proportion to their value as noted in Figure 4.12. Economic allocation distributes environmental flows in proportion to the cost of the products.

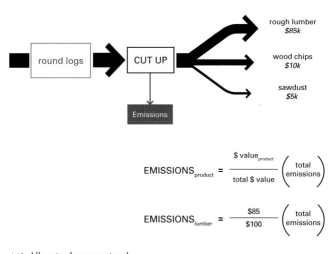

4.12 Allocation by economic value

Allocation can be a complex methodological challenge for LCA, one subject to debate among LCA practitioners and decision-makers alike. See Chapter 6 for more detail on challenges related to allocation methods.

4.7 Analysis: calculation methods

THE PRIMARY CALCULATIONS OF LCA are straightforward: simple multiplication, as outlined in Figure 4.13. However, the management of data can be quite cumbersome. Many LCAs are linear and the processes are independent of one another, enabling the calculations to be performed with simple multiplication. In more complex LCAs, the inventory flows can be interdependent.

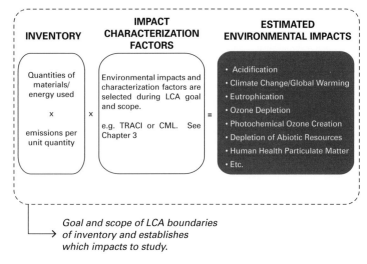

4.13 Simplified representation of primary LCA calculations

For example, in order to perform an LCA of steel that included the construction of the factory in the system boundary, some amount of steel would need to be purchased to make the factory for each volume of steel produced. This results in multi-variable dependent calculations that can be solved using matrix algebra (Heijungs and Suh, 2001).

One of the prime functions of LCA tools is to enable streamlined development of LCA models, tracking of data and calculating results. See Chapter 7 for more detail on types of LCA tools used in the building industry.

chapter 5

Life Cycle Assessment Standards

LIFE CYCLE ASSESSMENT STANDARDS are structured to provide a flexible methodology to enable LCA practitioners to customize the analysis as required to meet the goal and scope of the study. The ISO LCA standards 14040 and 14044 (ISO, 2006a, 2006b) are internationally recognized as the foundation standards for LCA, combining broad consensus and clear and adaptable guidance for performing an LCA. The LCA fundamentals presented in Chapter 2 outline LCA requirements defined by these standards. These standards explicitly state that "there is no single method for conducting LCA" (ISO 2006a: 9). They provide a framework to document the chosen LCA approach. ISO 14040 provides the overarching "Principles and Framework" of an LCA, defining key terms and outlining the four stages of an LCA described in more detail in Chapter 2. Additionally, it identifies "key features" of an LCA including such items as "LCA addresses potential environmental impacts; LCA does not predict absolute or precise environmental impacts …" (ISO 14040: 2006a, section 4.3: 9, with the permission of ANSI on behalf of ISO, © ISO 2013, all rights reserved) or "There is no scientific basis for reducing LCA results to a single overall score or number, since weighting requires value choices" (ISO 14040: 2006a, section 4.3: 9, with the permission of ANSI on behalf of ISO, © ISO 2013, all rights reserved).

ISO 14044 is formatted parallel to the structure of 14040 and provides more detail on the requirements for conducting an LCA. About twice as long as 14040, the ISO 14044 standard identifies both mandatory and desirable components of an LCA with a focus on documenting methodology and data choices to enable flexible application and accurate interpretation of LCA.

The ISO standards are broadly accepted yet the need to add detail and specificity to them is also recognized. Other more detailed LCA standards typically reference 14040 and 14044 and build on the fundamental principles outlined within them. These standards have been and are being developed by regional and international standards organizations such as:

- ASTM International Standards: (formally American Society for Testing and Materials);
- CEN: European Committee for Standardization;
- ANSI: American National Standards Institute.

Many standards reference the ISO 14000 series of standards to provide added detail to unify the LCA methods and reporting for buildings and building products. Figure 5.1 represents the relationship between these different standards and highlights that standards are constantly developing as regional and international consensus develops around methods of LCA appropriate for the building industry.

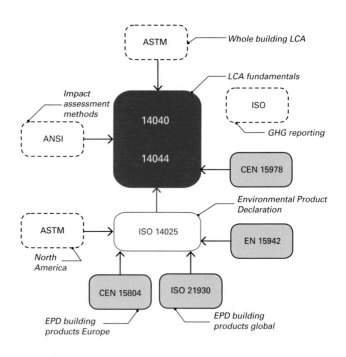

5.1 Interrelationship of global LCA standards

Table 5.1 lists the primary LCA standards published in Europe and North America. Other standards shown dashed in Figure 5.1 are currently under development or revision. This book provides an overview of topics covered by these LCA standards but should not be considered a proxy for these standards. Readers wishing to develop compliant LCAs and understand detailed requirements and terminology definitions should contact the standards organizations to obtain a copy of the current standards.

Table 5.1 Primary adopted LCA standards related to buildings and building products

Standard Number	Title	Reference
ISO 14040	Environmental management—Life cycle assessment—Principles and framework	(ISO, 2006a)
ISO 14044	Environmental management—Life cycle assessment—Requirements and guidelines	(ISO, 2006b)
ISO 14025	Environmental labels and declarations Type III environmental declarations Principles and procedures	(ISO, 2006c)
ISO 21930	Sustainability in building construction Environmental declaration of building products	(ISO, 2007)
CEN 15804	Sustainability of construction works – Environmental product declarations-Core rules for the product category of construction products	(CEN, 2012)
CEN 15942	Sustainability of construction works – Environmental product declarations Communication format business-to-business	(CEN, 2011a)
CEN 15978	Sustainability of construction works Assessment of environmental performance of buildings Calculation method	(CEN, 2011b)
PAS 2050	Specification for the assessment of the life cycle GHG emissions of goods and services	(BSI, 2011)
GHG Protocol	Product Life Cycle Accounting and Reporting Standard (GHG only)	(WRI/WBCSD, 2011)

5.1 Critical review

A FORMAL EVALUATION OF THE DATA, calculations and documentation by independent "third party" reviewers, known as a critical review, is required by ISO when the LCA is to be released publicly or used in a comparative assertion (see Section 5.2). The company that has commissioned the LCA typically organizes the critical review, identifying a chairperson to lead the review. The chairperson is charged with selecting other independent and qualified reviewers to assemble a committee of at least three members. This committee is charged by ISO 14044 (2006b: 31) to verify items such as:

- compliance with ISO standards;
- validity of the calculation methodology;
- appropriateness of the data used;
- that the interpretations reflect the stated goal;
- that the report is transparent and consistent.

The make-up of the committee is permitted to be quite diverse. In practice, the committee is typically hired by the company preparing the LCA and is composed of individuals with knowledge of the industry and/or LCA. The review can be internally verified (with the committee led by an employee of the company performing the LCA provided that individual was not involved with developing the LCA) or externally verified (completed by someone not employed by the company).

The review committee typically identifies questions and issues of concern regarding the LCA. The team developing the LCA can then either change the LCA to respond to the committee's comments or respond with additional clarification or justification of the LCA choices made. A report by the review panel is required that details the assessment of the LCA and identifies any concerns. All of the critical review panels concerns must be addressed, either by changing the LCA in response to the concern or rebutted/reported transparently within the final LCA report. In practice, most reviews undergo an iterative process until the critical review panel is satisfied with the final LCA and results and can write a simple letter confirming that the LCA confirms with ISO standards.

ISO 14044 (2006b) outlines in detail the sections of the LCA that must be evaluated. The names and affiliation of the critical review panel must be included in the final LCA. The reputation and affiliation of the reviewers thus will impact the credibility of the final LCA.

5.2 Comparative assertions

LCA RESULTS CAN ASSIST in making a "comparative assertion", stating that one option is environmentally preferable to another. If an LCA is being developed to compare products, this should be clearly stated as part of the LCA goal. Although many LCAs are conducted to compare different systems (such as studies to evaluate steel vs. concrete vs. wood construction), unless the studies follow explicit ISO requirements, the results cannot be used to assert that one method is preferable to another.

Using LCA results to make comparative assertions requires careful development of the LCA to ensure that the systems are functionally equivalent and that the data used is similarly precise, complete and representative of actual conditions. ISO 14044 requires that in addition to meeting the requirements for a critical review, additional reporting requirements are listed (2006b: 30–31). These reporting requirements are designed to ensure that an external "public" evaluator of the LCA can understand the data quality, underlying assumptions and potential weaknesses of the analysis used to declare a product or system as environmentally preferable.

In order to decrease the likelihood of misunderstandings or negative effects on external interested parties, a panel of interested parties shall conduct critical reviews on LCA studies where the results are intended to be used to support a comparative assertion intended to be disclosed to the public.

(ISO 14044: 2006b, section 6.1: 31, with the permission of ANSI on behalf of ISO, © ISO 2013, all rights reserved)

Comparative assertions require a critical review with more detailed requirements, including the broad requirement that "interested parties" should participate in the critical review (2006b: 6). One could interpret that to mean that competitors and environmentalists should be included in critical reviews of potentially controversial LCAs that establish comparative assertions. Interestingly, EPDs (see Section 5.3) can be used to establish comparative assertions, however, the EPD standards (ISO, 2006c; CEN, 2012) do not explicitly require such broad participation in PCR development and EPD verification.

The quantifiable metrics of LCA results are attractive for those looking to decide between two different options. However, the methods used to frame the LCA and the environmental impacts studied can alter results considerably. The debate between proponents of cloth and disposable diapers highlights this point. Different studies came to different conclusions. An evaluation of these different studies (Vizcarra et al., 2009) found that the results are inconclusive. If reducing water use is prioritized, disposable diapers are preferred. If reducing material consumption is prioritized, cloth diapers are preferred. Depending upon energy mix and modeling assumptions, the energy and emissions results were also inconclusive.

Similar results can be found in LCAs of steel and concrete commercial buildings. Depending on the methods used to model the structure, which LCA stages are included and the methods chosen to model allocation and recycling, either steel or concrete results in slightly higher LCA impacts. Many published LCAs comparing building products or systems have not undergone a comprehensive ISO critical review and thus technically should not be used to definitively state that one is superior to another.

5.3 Environmental labels: EPDs and more

THERE ARE MANY DIFFERENT "GREEN" PRODUCT-LABELING PROGRAMS. These can range from manufacturers' self-declaring green attributes (e.g. percentage of recycled content) in their packaging to third party verified declarations of LCA results (e.g. Environmental Product Declarations or EPDs) that meet established standards and criteria. There are hundreds of different eco-labels for different products and industries (Ecolabel Index, 2012).

ISO 14020 (2001) standardizes the general principles that all environmental labels and declarations should adhere to. These fundamental principles provide the foundation for more detailed standards related to environmental labels for products. The nine principles address issues such as: "Environmental labels and declarations shall be accurate, verifiable, relevant and not misleading …" (ISO 14020: 2001, section 4.2.1: 2, with the permission of ANSI on behalf of ISO, © ISO 2013, all rights reserved) and "Environmental labels and declarations shall be based on scientific methodology that is sufficiently thorough and comprehensive to support the claim and that produces results that are accurate and reproducible" (ISO 14020: 2001, section 4.4.1: 2, with the permission of ANSI on behalf of ISO, © ISO 2013, all rights reserved). ISO classifies environmental labels as Type I, II, or III.

Type I eco-label: certified "green"

Type I labels identify if a product meets multiple criteria of environmental performance based on LCA and/or other established scientific methods. Type 1 eco-labels have the advantage of being simple for consumers to interpret. A product either meets or does not meet the established standard as an environmentally preferable product.

ISO 14024 (1999a) provides standards for reporting these labels, requiring third party verification and outlining a structure for establishing the rules by which the data is calculated and outlining the performance thresholds that must be met. Type I eco-labels should fulfil the following conditions:

- are voluntary;
- are verifiable;
- meet criteria set by external party;
- are third party verified;
- address multiple environmental criteria.

An example of a Type I eco-label is shown in Figure 5.2. This "ECOLOCO", managed by UL Environment, is awarded to products and services demonstrating reduced environmental impacts. The product is evaluated from a life cycle perspective from raw material extraction, to production, distribution and disposal.

PRODUCT CERTIFIED FOR REDUCED ENVIRONMENTAL IMPACT. VIEW SPECIFIC ATTRIBUTES EVALUATED: UL.COM/EL
UL XXXX

5.2 Type I Label: ECOLOGO

(UL Environment ©, 2013 UL Environment (©2013) ECOLOGO *Product Certification*, Marietta, GA: UL Environment. Certification Program. Note: The UL ECOLOGO mark is a trademark of UL LCC. www.ul.com/el)

Type II eco-label: self-declared

Type II labels are single-attribute environmental claims addressing issues such as recycled content, energy consumption, etc. A company can define internal targets for environmental performance and identify and label their products accordingly.

ISO 14021 (1999b) defines key terms (e.g. "recyclable" and "reduced resource use") commonly used in environmental claims to enable comparable definitions. The standard has requirements to be followed for all claims as well as specific details for individual claims (e.g. "degradable" and "reduced energy consumption") and outlines procedures for internal evaluation and verification of claims. Type II eco-labels should fulfil the following conditions:

- be a single attribute;
- be internally verified;
- meet internally identified performance targets.

An example of a Type II label is shown in Figure 5.3. Type II labels are not common in the building industry. Likely the perceived value of third party verification and the wide variety of Type I labeling schemes available lead to less use of Type II labels.

5.3 Example of hypothetical ISO Type II eco-label

Type III eco-label: Environmental Product Declaration (EPD)

Environmental Product Declarations (EPDs) present LCA results and ISO 14025 (2006c) provides the baseline standards for EPDs. Critical components of an EPD are:

- are based on LCA data developed in accordance with ISO;
- are based upon established rules for conducting LCA;
- are verified by a third party "program operator".

As noted in earlier chapters, the results of an LCA depend upon the methodical choices made in developing the inventory and converting it to environmental impacts. EPDs result from LCAs that have been developed by a predefined standard set of rules (termed Product Category Rules, PCRs). Product categories are a defined product type or group of products (e.g. tile flooring or flooring) that can reasonably be developed under the same LCA rules.

Many different standards such as ISO 21930 (2007) and CEN 15804 (2012) have been and are being developed to provide the "core rules" for EPDs of building products. These standards provide guidance beyond the more general guidance provided by ISO 14044 and 14025, addressing requirements such as:

- who should be involved in developing PCRs;
- guidance on allocation methods;
- what to report in the EPD.

ISO 14025 was written presuming that most EPDs will be developed as business-to-business documentation but provides provisions to ensure that greater clarity is provided in business-to-consumer EPDs. A comprehensive EPD is difficult to interpret for those without extensive LCA experience. Although architects and engineers are likely to be considered "business" rather than "consumers", EPDs formatted for their use will benefit from more description and clarification on how to interpret the results.

Depending upon the requirements of the PCR and the EPD program operator, EPDs for building products can vary between simple declarations of total environmental impacts to detailed breakdown of impacts for individual life cycle stages. It is important to recognize that all PCRs are not the same, and some are more comprehensive and detailed than others. In order to meet the objective of being able to objectively compare products, the PCR must have enough detail to ensure that the LCAs are reproducible. This means that two different people would come up with close to the same results when performing an LCA in accordance with the PCR.

As LCA does not have established methods to report *all* environmental impacts, EPDs should include reporting of other known environmental issues. This poses an interesting challenge given the difficulty of identifying and agreeing upon which significant known environmental issues are appropriate to include in an EPD. An EPD program can help provide guidance and standardization across industries. Given that the US has many different program operators, this standardization is not yet occurring. The extent of other issues currently reported in EPDs ranges widely.

A comprehensive EPD should thus include a review and reporting of a wide range of potential environmental impacts ranging from human health, biodiversity and regionally specific environmental concerns. In current practice, most EPDs report only the typical LCA specific results.

CASE STUDY 5.1: EPDs OF THREE BUILDING PRODUCTS

This case study identifies three verified EPDs for building products to demonstrate the variety in format, detail and complexity that is possible. In addition to the summary data shown here, a typical EPD includes detailed information about the product, manufacturer and other information published as a requirement of the PCR. Figure 5.4 shows a portion of an EPD that reports total impacts with

	ATMOSPHERE		
	Global Warming Potential refers to long-term changes in global weather patterns – including temperature and precipitation – that are caused by increased concentrations of GHGs in the atmosphere.	**Ozone Depletion Potential** is the destruction of the stratospheric ozone layer, which shields the Earth from ultraviolet radiation that's harmful to life, caused by human-made air pollution.	**Photochemical Ozone Creation Potential** happens when sunlight reacts with hydrocarbons, nitrogen oxides, and volatile organic compounds, to produce a type of air pollution known as smog.
TRACI	6.79 kg CO_2-Equiv.	0.00000098 kg CFC 11-Equiv.	0.000015 5 kg NO_x Equiv.
CML	6.83 kg CO_2-Equiv.	0.00000094 kg R11-Equiv.	0.004 kg Ethene-Equiv.

FUNCTIONAL UNIT   One square meter of installed modular carpet for heavy duty wear
The reference flow is one square meter of modular carpet

5.4 Partial summary of results of a Type III EPD

(Courtesy of Interface Carpet. UL Environment, © 2013, *Interface Carpet EPD*, image is property of UL LCC. www.ul.com)

102

good information to help consumers interpret the data. In addition to reporting the environmental impacts using two different LCIA methodologies, the environmental mid-point impacts are explained to help the consumer assess the importance of the impact.

Figure 5.5 shows a straightforward "nutrition label" style of EPD developed for Armstrong ceiling tiles. This format reporting in a concise and consistent manner would help users compare between different products. Note that additional performance attributes such as reflectance and an ingredient list are included here.

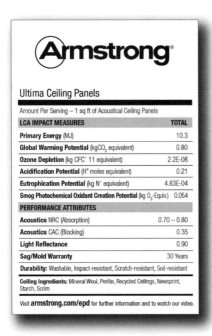

5.5 Summary results of Type III EPD

(Courtesy of Armstrong Industries)

Table 5.2 demonstrates a comprehensive EPD developed by the German EPD System, Institut Bauen und Umwelt e.V (IBU). This EPD provides detailed data broken down per life cycle stage appropriate for an LCA practitioner to integrate within a comprehensive whole building LCA and adapt as appropriate.

Table 5.2 Partial summary of results of EPD impacts

DESCRIPTION OF THE SYSTEM BOUNDARY (X = INCLUDED IN LCA; MND = MODULE NOT DECLARED)

Stage	Module	Description	Status
PRODUCT STAGE	A1	Raw material supply	X
	A2	Transport	X
	A3	Manufacturing	X
CONSTRUCTION PROCESS STAGE	A4	Transport	X
	A5	Construction-installation process	X
USE STAGE	B1	Use	X
	B2	Maintenance	X
	B3	Repair	MND
	B4	Replacement[1]	MND
	B5	Refurbishment[1]	MND
	B6	Operational energy use	MND
	B7	Operational water use	MND
END OF LIFE STAGE	C1	De-construction demolition	MND
	C2	Transport	X
	C3	Waste processing	X
	C4	Disposal	X
BENEFITS AND LOADS BEYOND THE SYSTEM BOUNDARYS	D	Re-use-Recovery-Recycling-potential	X

RESULTS OF THE LCA - ENVIRONMENTAL IMPACT: 1m² floorcovering

Parameter	Unit	A1-A3	A4	A5	B1	B2	C2	C3/1	C3/2	C4	C4/1	C4/2	D	D/1	D/2
GWP	kg CO_2-Eq.	8.82	0.185	0.556	0.003	0.29	0.01	0	0.029	9.83	8.3	0	-0.248	-2.56	-0.405
ODP	kg CFC11-Eq.	1.19E-12	3.3E-12	1.94E-8	0.0E+0	6.45E-9	1.8E-13	0.0E+0	2.6E-11	1.2E-6	7.2E-10	0.0E+0	-2.2E-10	-5.3E-10	-1.03E-7
AP	kg SO_2-Eq.	4.27E-2	8.42E-4	1.94E-3	0.0E+0	1.37E-3	4.68E-5	0.0E+0	1.39E-4	1.21E-3	5.47E-3	0.0E+0	-1.18E-3	-4.18E-3	-2.8E-3
EP	kg(PO_4)³-Eq.	1.35E-2	1.94E-4	7.15E-4	0.0E+0	1.99E-4	1.08E-5	0.0E+0	7.34E-6	5.65E-3	1.44E-3	0.0E+0	-6.19E-5	-3.45E-4	-5.82E-4
POCP	kg Ethen Eq.	3.09E-3	3.04E-4	1.11E-4	0.0E+0	1.79E-4	-1.69E-5	0.0E+0	8.21E-6	1.45E-3	3.91E-4	0.0E+0	-6.92E-5	-4.4E-4	-3.76E-4
ADPE	kg Sb Eq.	4.58E-6	6.88E-9	6.38E-6	0.0E+0	5.44E-7	3.8E-10	0.0E+0	4.05E-9	4.56E-8	1.57E-6	0.0E+0	-3.42E-8	-1.51E-7	-6.92E-8
ADPF	nvui	166	2.56	7.64	0	6.55	0.142	0	0.518	3.37	7.13	0	-4.37	-42.5	-68.1

Caption: GWP = Global warming potential; ODP = Depletion potential of the stratospheric ozone layer; AP = Acidification potential of land and water; EP = Eutrophication potential; POCP = Formation potential of tropospheric ozone photochemical oxidants; ADPE = Abiotic depletion potential for non fossil resources; ADPF = Abiotic depletion potential for fossil resources

RESULTS OF THE LCA - RESOURCE USE: 1m² floorcovering

Parameter	Unit	A1-A3	A4	A5	B1	B2	C2	C3/1	C3/2	C4	C4/1	C4/2	D	D/1	D/2
PERE	MJ	10.9	0.1	1.77	0	0.5	0.006		0.086	0.162	0.32	0	-0.729	-1.71	-0.183

(Courtesy of IBU)

5.4 Building industry-specific standards

GIVEN THE BROAD AND FLEXIBLE structure of the primary LCA standards, the development of more detailed guidance for use of LCA in the building industry is needed to help standardize methodological choices and improve comparability of LCAs.

Building industry-specific LCA standards have been developed and are developing in two primary areas: whole building LCA, and establishing core rules for building product EPDs. Figure 5.1 on p. 94 outlines current building industry-specific LCA standards to provide an idea of the range of standards work being undertaken simultaneously around the world. These standards are evolving quickly and readers should check with the standards organizations to attain current information.

Some of the aspects of LCA that are defined by building product EPD and PCR standards include:

- defining the stages of LCA;
- guidance to estimate the service life of products;
- allocation methodology;
- environmental impacts to report;
- information to include in the EPD.

The whole building LCA standards provide detailed requirements for items such as:

- the system boundary;
- building and component service life;
- environmental impacts to report.

As these standards develop and are more widely adopted, the ability to compare between LCA studies will be greatly increased. Additionally, the standardization process provides a clear forum to debate different perspectives and come to consensus upon how best to deal with some of the methodical challenges identified in Chapter 6.

chapter 6

Methodical Challenges

LCA ACCOUNTS FOR AND EVALUATES the inputs from nature and the outputs to nature and potential environmental impacts of a product system throughout its life cycle. The results can vary based on methodical choices made by the LCA practitioner. For this reason, users of LCA data must be very careful when comparing results from different LCA studies or different LCA data sources as the choices made in conducting the LCAs may be so different as to preclude comparison.

Some of the typical LCA questions that require professional judgement and interpretation include:

■ If a manufacturing plant emits 100kg of CO_2e and produces two different products, what percentage of the emissions should go to each product?
■ If the "waste" of product manufacturing can be used in making another product, is it really a "waste" or is it "secondary materials and fuels" (a co-product)?
■ Is industry-average data adequate to use when comparing different products?
■ How significant is it that we do not know the actual life span of a building?

This chapter presents case studies of methodical challenges related to allocation and uncertainty that can impact LCA results.

6.1 Allocation: by-product, co-product or waste?

WHEN A MANUFACTURING SYSTEM PRODUCES more than one product, the inputs and emission are typically proportioned to each of the different products using a specific allocation method (see Section 4.6). Deciding what that proportion should be can be challenging. Although LCA standards provide guidance on appropriate allocation procedures, judgement is still required to make sure that the LCA choices support the goals of the LCA study (ILCD, 2010a: 275). Different perspectives, scientific logic and government policy can influence which allocation method is deemed "preferable".

CASE STUDY 6.1: TIRE END-OF-LIFE ALLOCATION QUESTION

Tires are manufactured for use in driving automobiles. Used tires have one of two typical end-of-life states: they are either landfilled or used as fuel for industrial processes such as cement kilns. When a tire is used as fuel, should the emissions related to the combustion of a tire go to the car or the cement kiln?

Five different arguments for the appropriate allocations of the emissions resulting from tire combustion in a cement kiln are given below.

ARGUMENT 1

In the US, most tires are landfilled and thus burning the tires to produce cement results in faster and more certain emissions to air. The tires are not fully used. Thus, the emissions related to use as a secondary fuel should be allocated to the product of cement. The car gets neither a credit (for generating "secondary" fuel that avoids primary fuel use) or a debit (for tire disposal to landfill). The emissions related to transporting the tire to the cement plant gate and combusting the tire are allocated to cement (Figure 6.1).

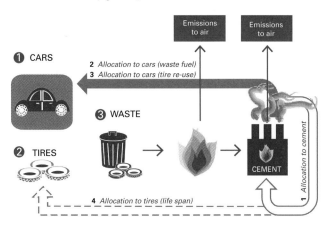

6.1 Allocation options for emissions of tire used as fuel for cement manufacturing

Environmental burden of tire = Production of tire + Waste treatment

Environmental burden of cement = Production of cement + Treatment of tires − Avoided landfill of tires

ARGUMENT 2

We are running short on fossil fuels. We should promote the use of waste fuel sources by allocating the emissions to the car. This would reduce the environmental burden to cement and encourage use of waste as fuel. The energy that is avoided because the tires are used as fuel can be accounted for by expanding the system boundary and providing emission credits to the car/tires. In this case the environmental allocation would change.

Environmental burden of tire = Production of tire + Treatment of tire for fuel use − Avoided production of energy

Environmental burden of cement = Production of cement + Avoided energy production

ARGUMENT 3

All of the supply chain sources including tire, car and cement manufacturers should work to reduce emissions and thus the emissions from combusting tires should be allocated 100 per cent to each of the three industries. This method is called "over-counting" and would not be an internally consistent method of LCA.

LCA methods such as those outlined in the *ILCD Handbook* (ILCD, 2010a; CEN, 2012: Annex B) provide clear guidance on how to determine if the product (a service, product or material) of a manufacturing process is a co-product (interrelated or interacting activities per ISO, 2006b) or a waste (item which the holder disposes of per ISO, 2006b). However, this clear distinction can be muddled by government policies defining products as "waste" independent of their value.

CASE STUDY 6.2: SLAG – WASTE OR BY-PRODUCT?

In a steel furnace, iron ore is heated up and smelted, which separates the desired metal from the other elements within the raw material. These impurities, termed slag, can be removed from the molten metal. Slag can form granules during the cooling process. The slag can be landfilled or used as roadway base, fertilizer or a cementitous material in concrete production. Depending upon its chemical composition, ground slag can act as a binder to supplement the cement used in concrete.

To be useful as a supplementary cementitious material, the slag must be ground and processed. Cementitious slag is a commodity that has value. Based on LCA standards noted above, the slag would be considered a co-product and some of the emissions related to producing the steel would be allocated to the slag. According to the US EPA, slag is classified as a waste material. European legislation (EU, 2008) mandates allocation of emissions for any co-product with economic value. Thus, in Europe, ground granulated blast furnace slag (GGBFS) would likely report a higher environmental impact than would be reported in the US.

The goal of the life cycle analysis as well as the impact of the methodological choice should be evaluated when assessing which method is more appropriate. A strong argument could be made to support either allocation method in the case of slag used in concrete (Figure 6.2).

ARGUMENT FOR WASTE

In an LCA standard for building products (CEN 15804, 2012: 28), the section on co-product allocation clarifies that the purpose of the plant should be considered when assessing allocation methods. In this example, the steel mill is designed to be a steel mill. In the US, a significant amount of slag is still being landfilled and thus treating slag as a waste product seems appropriate. The proportion of the slag that is landfilled, in any region, should be considered as waste and allocated to the product, in this case, steel.

ARGUMENT FOR CO-PRODUCT

An equitable distribution of environmental impacts would require that the total life cycle impacts be proportioned between the primary product, steel, and the by-product, slag and other steel industry by-products. Interestingly, different studies using economic allocation have resulted in different conclusions.

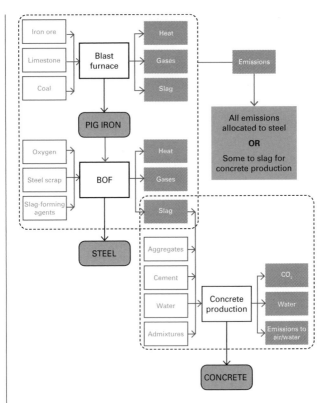

6.2 Allocation options for emissions related to slag production

6.2 Allocation: recycling methodology

ALTHOUGH THE BENEFITS OF RECYCLING may seem obvious, choosing how to model the impacts and benefits of recycling in LCA can be complex. Varying how recycling is modeled can significantly change the LCA results. There are two approaches most commonly used to model recycling: (1) the recycled content method (focusing benefit on the use of the recycled material); and (2) the end-of-life recycling method (focusing benefit on how much material is recycled at the end of life). This promotes the use of materials that do not have a well-established recycling practice and promotes the recyclability and collection of recyclable materials. These two methods are based on different system boundary assumptions.

Recycling that occurs internal to the manufacturing facility (e.g. scrap steel that is recycled within a factory) is termed closed loop product system. Recycling that occurs outside of the manufacturing facility (e.g. use of recycled aluminium from window walls in aluminium can production or tires recycled into rubber flooring) is typically termed an open loop product system.

Closed loop recycling allocation can be either a closed loop product system or an open loop system in which the scrap material can be re-used while maintaining the existing properties (e.g. structural steel can be recycled to be used again as structural steel).

Open loop recycling allocation occurs when the material properties are changed for the next application (e.g. when concrete is converted to roadway base after recycling).

As shown in Figure 6.3, in a closed loop system, the re-use of scrap avoids the need for additional raw materials. All the energy needed to process and utilize the scrap would be captured in this analysis so the reduced impact through the use of scrap would be captured directly through the reduced need for input raw materials. Figure 6.4 represents an open loop system in which recycled scrap is input to the system and recycled steel is output from the system but the recycling loop is "open" as the steps between end of life and

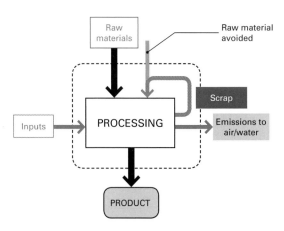

6.3 Closed loop product system

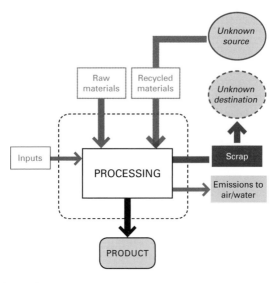

6.4 Open loop product system

scrap input are not defined, tracked or often even known. In reality, many manufacturing systems include both open and closed loop recycling. Figure 6.5 shows the system boundary of a system that both creates and uses scrap steel and internally recycles steel. This product system has both closed and open loop recycling taking place, however, the LCA should use a closed loop recycling system as the steel maintains its properties in both recycling cases.

An open loop system includes scrap as a "co-product" and a material input and thus the LCA must address the allocation of the emissions related to the input of recycled materials and the emissions, or the emissions avoided, related to the production of useable scrap.

Figure 6.5 shows a simplified system boundary for creating reinforcing steel. Within the cradle-to-gate system boundary of the steel product, internal closed loop recycling takes place. Within the cradle-to-grave system boundary, some of the product is recycled at end of life to an unknown destination, an open loop system. In this case of reinforcing steel, the manufacturing of the rebar consumes a large amount of scrap steel from an unknown external source. At the end of life, some of the steel will be recycled as scrap and some will be left *in situ*.

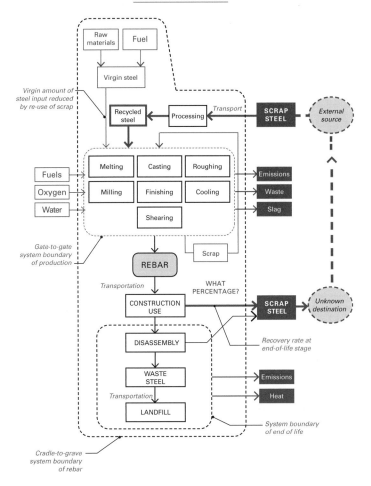

6.5 Cradle-to-grave system boundary of steel reinforcing bar with recycled content allocation method

THE RECYCLED CONTENT METHOD assigns the impacts from extraction, refining, manufacturing and demolition of recycled metal to the first use of the material. The impacts from transportation, processing and re-use of the recycled metal are assigned to the second product. Thus the environmental impact of materials with high percentage of recycled content is less than that of virgin metal. Recycling at end of life is not modeled as the end-of-life impacts of disposing metal are negligible and no credit is given for the amount of metal recycled. Thus the recycling rates of a product will not impact the LCA results. The system boundary shown in Figure 6.5 represents the recycled content method: scrap steel comes with zero "burden" from the initial manufacturing of the steel, the scrap itself is outside of the system boundary.

This method would favour the use of recycled material and would lead to a focus on increasing the use of recycled materials and would be useful in promoting the recycling of materials. The International Council on Mining & Metals (Atherton, 2007) notes that metal recycling is economical and the market is "mature" and thus a policy to reward recycling is not necessary and might not result in improved performance. This approach does not help decision-makers better manage metal use based on actual environmental performance.

THE END-OF-LIFE APPROACH focuses on how much metal can be recycled at the end of a product cycle and thus replaces the need for the production of primary material. The avoided production of primary material is then credited to the system at the end-of-life stage for products that result in recyclable steel. Expanding the system (see Section 4.6), to include the avoided production processes to credit the process creates the recyclable material (a credit for the net scrap production and consumption).

The end-of-life allocation method would reward products that are highly recyclable and not directly incentivize the use of recycled material. The International Council on Mining & Metals recommends this method because it "encourages manufacturers, policy-makers and other decision-makers to evaluate real performance and improve the design and management of products, including their disposal and recycling" (Atherton, 2007). Note the cradle-to-gate LCI results of EOL recycling of different steel products with different input amounts of scrap would not vary based upon the input amount of recycled material used but would vary based upon the amount of product that is recycled at the end of its life.

The World Steel Association *Life Cycle Assessment Methodology Report* (WSA, 2011) outlines calculation methods that assign credits related to the production of scrap equal to the environmental "costs" assigned to the use of scrap. Using industry-average data, the avoided burden of production can be computed.

Figure 6.6 represents a comparison between two steel manufacturing systems: Product 1 (high recycled content and low recycling rate) and

Product 2 (from a mill using less recycled content and a product with a high recycling rate). Figure 6.7 demonstrates how the choice of recycling allocation method could impact results. The World Steel LCI methodology report outlines these calculations in detail (WSA, 2011: 76). Note the net impacts are lowest for Product 1, the high recycled content product, using the recycled content method. For Product 2, the end-of-life method produces lower results, given the high percentage of scrap that is recovered for recycling.

The World Steel Association LCI recommends applying the end of life/closed loop recycling methodology and the LCI data for global and European steel is reported based upon this method but also provides cradle-to-gate data. The rationale for this (ibid.: 73) includes:

■ scrap has economic value: no need to create demand;
■ quantity recycled is driven by end-of-life recycling rates;
■ demand for scrap exceeds the availability.

This method would also "level the playing-field" between producers of basic oxygen furnace (BOF) and electric arc furnace (EAF) mills. As noted, data produced on these different methodologies is not comparable and thus interpreting LCI results requires an understanding of the recycling allocation methods used. LCI data in the US typically reports using the recycled content method, prioritizing materials with high recycled content.

Figure 6.8 captures the complex flow of steel materials through the manufacturing process for the UK. There are two main types of mills that produce steel. The BOF uses a high percentage of virgin steel. The EAF uses a high percentage of recycled scrap steel but requires the addition of pig iron/virgin steel. As shown in Figure 6.8, different materials are typically made from the different manufacturing processes. This results from both the capabilities of the mills (e.g. BOFs have more ability to create specific steel properties, given the more consistent input of raw materials) and also regional manufacturing variability (e.g. EAF "mini-mills" increased production in US during the late twentieth century when larger BOFs were struggling to maintain global competitiveness). Distributions for other regions will be different but the complexity of material flow will be similar.

The manufacturing processes for these two methods of steel production are significantly different, depending upon different fuel sources and using different proportions of recycled content. Thus, the average life cycle impacts for steel will depend upon the manufacturing and fuel mix of a region. Steel produced in an EAF powered by hydropower would have a relatively low environmental footprint compared to steel produced in a BOF with relatively dirty coal.

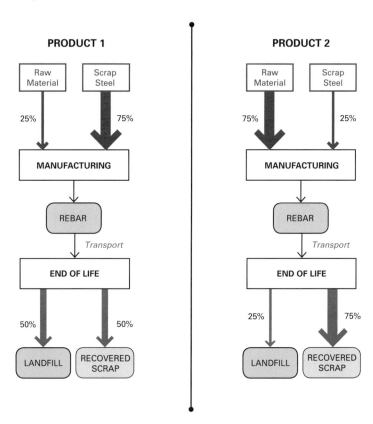

6.6 Two hypothetical steel manufacturing processes

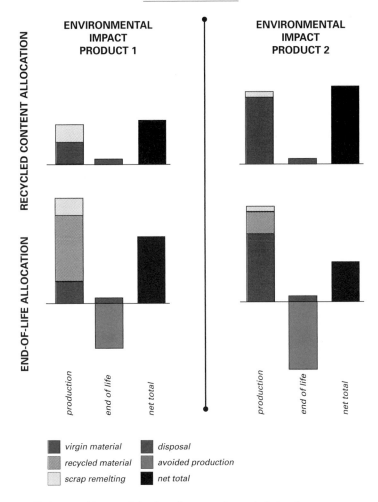

6.7 Environmental impact variation depending on recycling method selected
(Adapted from World Steel (WSA, 2011: 70–71)

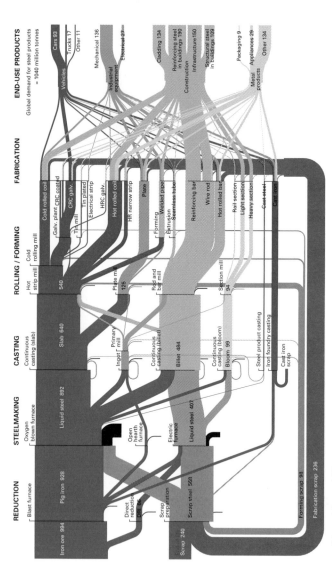

6.8 Material flow in UK

(Adapted from Cullen *et al.*, 2012)

6.3 Allocation: biologically-based carbon

ACCOUNTING FOR THE CARBON ABSORBED during growth of plants can make for an interesting challenge in interpreting the LCA results of bio-based products such as lumber. Looking at the issue simply, carbon dioxide is removed from the atmosphere and converted into sugars that are stored in leaves, stems, trunks and roots of trees and other plants (Figure 6.9).

Forests play a huge role in the global carbon balance. They absorb carbon during growth and emit carbon (and other chemicals and materials)

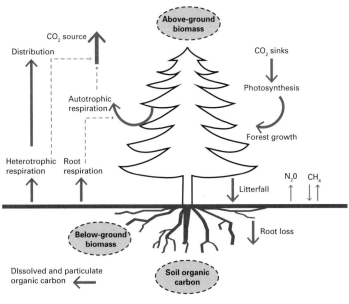

Carbon stocks

- Above-ground biomass
  - stemwood
  - branchwood
  - berk
  - foliage
  - seeds
- Below-ground biomass
  - coarse roots
  - fine roots
  - stumps
- Litter
- Coarse woody debris
- Soil organic carbon

6.9 Carbon absorption during plant growth

(Adapted from Morison et al., 2012 with permission from the Forestry Commission) when wood decomposes or burns. Changing tracts of land from forests to agricultural or other uses can impact both local and global environments. This section provides an overview of issues related to biologically-based carbon in LCA.

Note that the typical LCI of wood products does not track the environmental impacts of respiration, combustion, leaching and erosion. The fuel and fertilizer used during forest planting, management and harvest are the major LCA inputs. Conventional LCA does not capture the environmental aspects of forest management (such as habitat, water quality and social impacts) that forest certification programs address. The life cycle of a forest product can be divided up into the stages of growth, manufacturing, use and end of life as shown in Figure 6.10.

The details of the growth and harvesting phases of the LCI outlined here are adapted from a published LCA (Puettmann *et al.*, 2013). The boundary of a typical forest growth LCI includes: site preparation, seedling growth (fertilizer, energy and water), planting the seedlings and forest management (site preparation, thinning, and fertilization). Timber harvesting activities include five components: felling (severing the standing tree from the stump), processing (removal of limbs and tops and cutting of the tree into sellable and

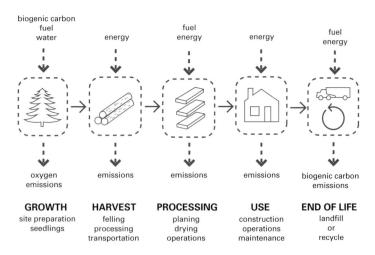

6.10 Life cycle stages of a wood product

transportable log lengths), secondary transportation (called skidding or yarding, which is moving trees or logs to a loading point near a haul road), loading (moving logs from the ground to haulage vehicles) and transportation (typically by truck) to the lumber mills. The LCI tracks fuel use and equipment type and efficiency to determine combustion emissions.

A forest can act as a carbon sink when the amount of carbon taken in and stored by the soil, trees (and resulting timber products) and other forest vegetation is greater than the total amount of carbon dioxide emitted due to respiration, decay, disturbances (such as harvests or fires) and emissions due to wood processing.

Figure 6.11 tracks the carbon of a wood product originating from a single tree over time. Over the full life of a wood product, the wood can only emit as much carbon as it originally absorbed and can be considered "carbon-neutral". However, the carbon and other emissions related to processing and manufacturing must still be accounted for. Net carbon emissions increase during planting, manufacturing and end of life, and decrease during growth when carbon is absorbed.

The carbon that is stored over the full life of a wood product is keeping CO_2 out of the atmosphere. There are different methods for tracking and reporting this long-term wood product carbon storage. At the national or global level, the wood product carbon storage pool can be reported by

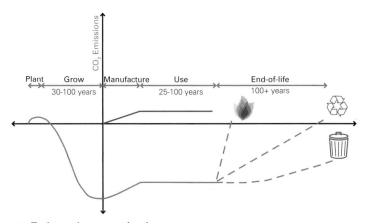

6.11 Tracking carbon in a wood product

(Adapted by permission from work at Arup)

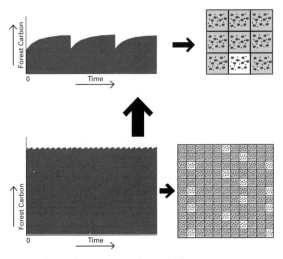

6.12 Tracking carbon over time in forests of different size

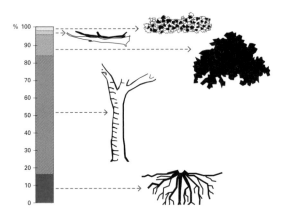

6.13 Carbon distribution within a typical tree

calculating the annual change in wood products in use and in landfill or the "harvested wood carbon" (EPA, 2013g).

The red line in Figure 6.11 indicates the carbon impacts as reported in US LCI databases (Puettman *et al.*, 2013), reporting the emissions from manufacturing only and treating the absorbed and released biogenic carbon as net neutral. This method is conservative only if trees are planted to replace those that are harvested, as it does not take into account the time delay of the carbon sequestration that long-life products can provide.

Although tracking the carbon over time from a single tree can help us to understand the relationships between biogenic carbon and the processing emissions in wood products, this does not reflect the more dynamic nature of forest growth. Provided trees are planted to replace the trees that are harvested, the total forest carbon can be modeled as constant. If forests were not managed sustainably (harvests greater than replanting), the net forest carbon would decrease. Thus one can evaluate forest carbon over time focusing upon a single tree, a plot, a forest, region or the world. Figure 6.12 represents how the single line representing carbon of a tree shown in Figure 6.11 can become more constant when the carbon of larger forest areas are evaluated with multiple plots of land harvested and planted at different time frames.

The carbon stored in wood ranges between 46 and 55 per cent (Lamlom and Savidge, 2003) of the total weight. Thus, a tree that weighs 1000 kg (oven dry) would contain approximately 500 kg of carbon. Figure 6.13 presents the distribution of carbon in a typical tree. Note that the majority of the carbon is stored in the trunk.

Within a typical tree, the branches, and litter, constituting about 15 per cent of the total carbon content, are either left in the forest to rot, are burnt in waste piles, or burnt as a bio-fuel. As most LCI data does not track biogenic carbon, the differences in these forest management practices are not captured by conventional LCA. The roots are typically left in the ground to rot. The root mass of a managed forest is assumed to be carbon-neutral as well, assuming that the roots rot at approximately the same rate as they grow.

Standing carbon represents an estimate of carbon in the standing forest based on the estimate of biomass. The rates of decomposition of downed material, down foliage, and/or roots from harvested trees vary and different studies reach different conclusions (Prescott, 2010). Johnson et al., (2005) concluded that "In both the South-east and North-west of the United States … fine roots grow and decompose at about the same rate". This means they do not add net carbon to the system. These models predict that both native forests and "sustainably managed" forests (replanting after cutting) are carbon sinks, provided that the wood is used in products that outlast the growth cycle.

6.4 Uncertainty: data variability in product LCA

MOST PUBLISHED LCI data represents industry-average data. As noted in Section 4.5, uncertainty can be introduced by randomness, choice or lack of knowledge or experience during the development of the inventory of materials and energy, by the use of uncertain LCI data sources or by the errors in development of the analysis model. Some uncertainty is quantifiable and can be reported as known variability, such as the range of efficiencies reported in an industry survey that is input into an industry average value. Other uncertainty is not known in sufficient detail to be quantified.

When using LCA results to compare options, understanding the precision of the result is critical to understand whether or not there is a statistical difference between two options. See Figure 6.14 for an example of the carbon footprint of five different products. Each performs the same function but has significantly different average carbon footprints. Based on average data alone, one might say that product E is significantly better than product A. However, given that the range of expected values is quite high, one might conclude that the difference between products is less significant than first envisioned.

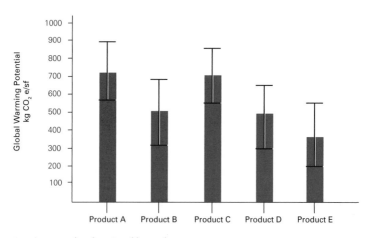

6.14 Average carbon footprint of five products

CASE STUDY 6.3: SOURCES OF UNCERTAINTY IN CONCRETE

Table 6.1 demonstrates potential sources of uncertainty that could develop when conducting an LCA of ready mixed concrete. Some of this variability could be quantified while others are unknown.

Table 6.1 Potential sources of uncertainty and variability in concrete LCA

LCA Stage	Random	Choice	Ignorance
Activity Data	Cement fuel source quantities fluctuate depending on market conditions.	Using company average data for each of the individual concrete mix designs.	Unaware that a maintenance process consumes significant resources.
LCI Data	Variation in CO_2 emissions from cement chemical reaction due to natural variety of input materials.	Using LCI data for European admixtures for US concrete LCA.	Unaware that published LCI data is gate-to-gate rather than cradle-to-gate.
Model	Average cement data based on old plant surveys used to predict uncertain future condition.	Excluding processes from the system boundary.	Unaware that the LCA model didn't accurately account for transportation "back-haul" as trucks return to source.

As architects increasingly look to LCA to help in assessing the relative potential environmental impacts of materials and processes, the estimated uncertainty and known variability should be included in published LCA results so that users do not get an inaccurate perception of the level of precision of LCA results.

6.5 Uncertainty: variability in whole building LCA

AS NOTED EARLIER, BUILDINGS CAN BE CONSIDERED PRODUCTS: large ones that have a long life span. Depending upon when during the design, construction or use stages of a buildings life the LCA is performed, there will inherently be greater or less uncertainty about the inventory of materials and energy used. However, there remains uncertainty related to the LCI data to use and modeling methodologies.

CASE STUDY 6.4: WHOLE BUILDING LCA UNCERTAINTY

Table 6.2 demonstrates potential sources of uncertainty that could develop when conducting an LCA of a commercial office building.

Table 6.2 Potential sources of uncertainty and variability in a whole building LCA

LCA Stage	Random	Choice	Ignorance
Activity Data	Variation in the occupants' use of energy.	Defining a future scenario such as building life span.	Unaware of a maintenance regime.
LCI Data	Variation in modes of transportation used to deliver materials.	Selecting industry average LCI rather than process specific LCI.	Unaware that there are multiple technologies to create similar products.
Model	Modeling assumptions regarding energy use.	Excluding a process from the system boundary.	Omitting unknown construction activities.

As noted throughout the text, there are many sources of variability when performing an LCA: system boundary, data sources, allocation methods, etc. Two different individuals in seven different iterations analyzed the embodied impacts of a medium-sized office building and the breakdown of the calculated global warming potential impacts is shown in Figure 6.15.

The primary sources of variability in these analysis occurring in all three LCA stages were related to choices made by the researcher: differences in assumptions of material quantities, different sources for LCI data selected, and different scopes of assemblies included in the system boundary.

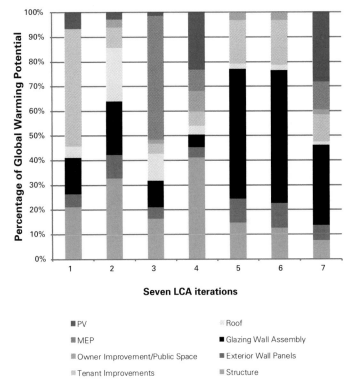

6.15 Seven LCA iterations demonstrating variability in results depending on modeling methods

Understanding the percentage of impacts attributed to the materials and products used in construction relative to the impacts related to operating enables architects and policy-makers to evaluate where best to focus their efforts. Figure 6.16 plots the embodied impact as a percentage of the total (operating plus embodied) for a hypothetical building in which the design options (such as configuration, glazing area, construction material, etc.) have been modeled parametrically to test 5,000 iterations of potential configuration (Basbagill, 2013). For the majority of the iterations the embodied impacts were less than 15 per cent of the total impacts. While this study reinforces that operational impacts are generally dominant, it also highlights

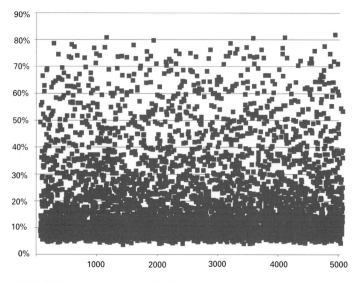

6.16 Embodied impacts as a percentage of total impact

(Basbagill, 2013)

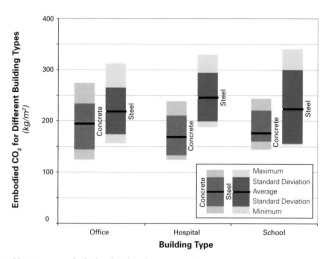

6.17 Variations in embodied carbon based on structural system

(Adapted from Arup, 2012)

that the variability is large and that each building case should be evaluated individually.

Another LCA study performed by Arup International, comparing different structural systems (Arup, 2012), found that the variations within structural types of concrete and steel structures could be more significant than the differences between systems. See Figure 6.17 on page 128 for a summary of their findings.

6.6 Uncertainty: end-of-life assumptions

ESTIMATING THE END-OF-LIFE IMPACTS is particularly challenging for long-life products such as buildings as the uncertainty as to when and how a building's life span will be terminated is large. In many cases an LCA scope includes estimates of future life cycle stages such as use and end of life. In this case future scenarios must be defined. The LCA can be conducted for one specific scenario or several variations on the scenarios can be evaluated to understand the range of options.

The European Standard CEN 15978, *Sustainability of Construction Works: Assessment of Environmental Performance of Buildings* (2011b) provides detailed guidance on how to define and analyze scenarios based on EU policy.

Current waste treatment methods can be reported, as shown in Figure 6.18 on page 130, or through industry surveys. These models can be used as a proxy for future conditions, assuming recycling and combustion rates remain unchanged. Alternatively, the LCA could evaluate trends in waste treatment and predict conditions at a future date.

Establishing the appropriate end of life of a whole building and its components thus requires an end-of-life decision tree to be evaluated, such as is shown in Figure 6.19 on page 131. The choices made in this process can have a significant impact on the LCA results. In some LCA models (Athena, 2012), the end-of-life stage system boundary only includes the transportation of materials to a landfill.

LCA studies (DEQ, 2010; JRC, 2011) provide additional guidance on waste treatment modeling and designing for end of life. Wood products that are incinerated for energy release the majority of their biologically absorbed carbon at combustion but can be modeled to credit the avoided impacts that would have been emitted from the combustion of fossil fuels.

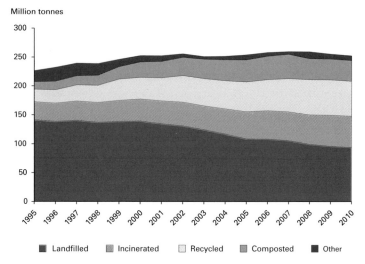

6.18 Municipal waste management in Europe, 1995–2010
(Adapted from EEA, 2012: 104)

As modeling future scenarios for long-life (and unique) products such as buildings is significantly more difficult than for a short-life and mass-produced product such as a beverage can, great care should be used when integrating end-of-life LCA impacts into whole building analysis. Even efforts to plan for end of life such as designing for deconstruction, must be evaluated, acknowledging the uncertainty inherent in these predictions.

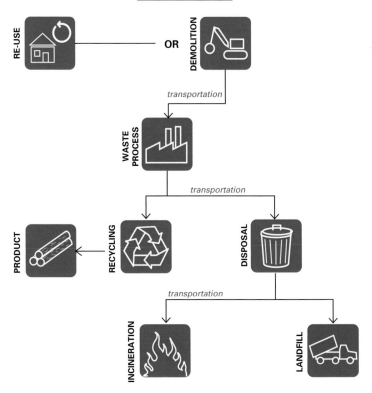

6.19 End-of-life decision tree

chapter 7

Implementing Life Cycle Assessment

FOR THOSE WISHING TO REDUCE the environmental impact of buildings and who prefer quantifiable metrics to generalized recommendations, LCA is an attractive method. However, the optimism of initial opportunity can be dampened by the challenges of managing data and interpreting the complex issues such as by-product allocation. In some instances a relatively simple analysis to interesting questions (e.g. is it better to drive a further distance to deliver a lower environmental impact product or pick a higher impact product that is closer?) provides satisfactory answers. In other instances, complex analysis, performed by expensive consultants attempting to answer what appears to be simple questions (e.g. which has a lower environmental impact, steel or concrete commercial buildings?), can deliver inconclusive results.

As LCA becomes more understood and integrated into building design and construction practice, the value, accuracy and effectiveness of LCA should continue to be evaluated.

VALUE:
What are the actual costs and benefits of the analysis?

ACCURACY:
Are the results accurate and/or precise enough to make definitive recommendations?

EFFECTIVENESS:
Does LCA result in a substantial reduction of impacts? Could other (more prescriptive methods) provide similar results with reduced efforts?

The implementation of LCA in the building industry is an emerging form of practice. There is significant opportunity for interested individuals to link the data and methods of LCA to building design and construction practice to enable reductions in the environmental impact of building construction and operation. While doing so, the optimism of LCA potential should be tempered by an understanding of the limitations and challenges. Section 7.1 outlines credible optimistic and pessimistic views of the future of LCA in the building industry and its integration to develop a more tempered view of the opportunities and challenges ahead. Section 7.2 provides resources for those looking to implement LCA in their practice.

7.1 LCA: optimism, pessimism and perspective

IF ONE IS OPTIMISTIC and assumes that the current challenges of LCA are just growing pains, the future of LCA in building design and construction practice is bright. The expectant view assumes that key developments such as the following will occur:

- Government and industry will collaborate effectively to develop comprehensive LCI datasets with consistent reporting methods.
- Manufacturers will integrate LCA into their practices, reporting LCA data in EPDs and using LCA results to inform and improve their manufacturing practices.
- LCA tools will be developed for the building industry to integrate environmental assessment into the building design process.

With fully integrated LCA data and tools, building LCAs could be developed with increasing specificity and could enable real-time decision-making as environmental impacts can be assessed throughout the whole design process. As shown in Figure 7.1, the product LCA data can feed into a whole building assessment.

7.1 Product LCA informing whole building LCA

The environmental impacts of LCA could be linked to the economic models of construction and operating costs to create assessment tools that enable the optimization of multiple performance criteria. Rather than requiring LCA studies as discrete analysis, LCA could be embedded into current processes such as developing three-dimensional models for construction or preparing construction cost estimates. These models should enable a growing level of detail and specificity as projects move from conceptual ideas to defined projects, enabling the scale and precision of LCA to evolve with the project.

At the time of writing, tools that link the software of building information models (BIM) and LCA databases are under development (e.g. KieranTimberlake, 2013). As these tools are refined to enable the integration of product-specific EPD data and include both industry-average defaults and user over-ride on options such as transportation distances and manufacturing processes, the production of whole building LCA will be significantly easier and standardized. The integration of user input and LCA methods and the data required to enable consistent LCA results is represented in Figure 7.2.

Governmental efforts, such as in France (HQE, 2012) and the NGO policy, such as the inclusion of LCA in green building codes and rating systems (ICC, 2012; USGBC, 2013) promoting and/or rewarding the use of both EPDs and whole building LCA, are spurring interest in LCA from manufacturers and designers. This increased market demand will likely help drive the needed advancement in LCA data and tools. Increased use of LCA by building industry professionals will help advance a more holistic, life cycle

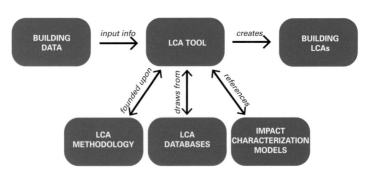

7.2 Integrated LCA framework

approach to buildings, helping designers and policy-makers move beyond their fixation with fixed costs and encourage them to include environmental impacts and total life cycle impacts when making building and construction choices.

PESSIMISTICALLY, ARGUABLY, LCA IS BEING OVERSOLD. A huge effort is required to conduct an LCA and the results can be of questionable value: uncertain with limited ability to impact change. One can argue that even in the relatively certain field of operational energy use (with meters to track consumption), we find that our ability to predict actual energy use has limited precision. How can we expect to develop credible data throughout a complex and changing supply chain and the uncertainty of building life cycle stages?

Given the complexity of a comprehensive LCA, some believe that methods to reduce the environmental impact of buildings can be generalized from LCA studies and do not require product or building-specific LCAs. Some recommendations that might be extrapolated from these studies include:

■ consume less energy;
■ build smaller buildings;
■ re-use existing buildings and materials;
■ design efficient structural systems;
■ design for longevity of materials and buildings;
■ use low embodied energy wood from sustainably managed forests.

These recommendations could be founded on robust LCA studies without requiring individual designers or manufacturers to conduct detailed LCAs. Unless LCA can become cost-effective and integrated within existing design and manufacturing processes, the effort of conducting an LCA can often outweigh the perceived benefits.

TEMPERING OPTIMISM WITH PRACTICALITY, the integration of LCA methods and data with design, construction and manufacturing practices provides opportunities to increase knowledge, shape policy and practice, and improve practice and manufacturing methods to reduce the environmental impact of buildings and building products. The limitations and requirements of LCA outlined in earlier chapters should be thoroughly understood so that the analysis is appropriate to desired application. Tools, data and standards are being developed that support more effective integration of LCA into building practice. Current opportunities depend upon the perspective and goal of the LCA study and can range from actionable activities to exploratory research.

Actions for design and construction professionals

- Specify products that report LCA results using EPDs and prioritize low impact products.
- Specify concrete environmental performance/LCA results, in addition to the typical performance measures of strength and workability.
- Conduct LCA studies of buildings and/or building components, enabling the inclusion of environmental impacts within a decision-making framework.
- Use LCA principles to conduct and report the corporate carbon footprint.
- Develop an internal database of projected LCA results/carbon footprints to better understand trends and opportunities in individual market sectors.
- Participate in LCA standards development, particularly PCR creation.
- Advance a life cycle approach to buildings, enabling items such as:

    a) better planning for maintenance and end of life;
    b) the ability to articulate the value of planning for life cycle vs. first costs;
    c) the integration of environmental performance as an additional rationale for prioritizing resiliency (e.g. improved seismic performance).

Actions for manufacturers

- Conduct an LCA of products and/or manufacturing lines.
- Use LCA results to identify opportunities for improvement and establish performance improvement targets.
- Develop EPDs for products.
- Participate in LCA standards development, particularly PCR creation.
- Collaborate with industry trade organizations to develop industry-average LCI data.

Research needs

- Better quality open access LCI data for building materials and products.
- Improved LCI and LCA reporting of variability and uncertainty.

- Establishing if "benchmarking" embodied impacts of buildings is practical.
- Improved practice guides for implementing LCA for building products and whole building analysis.

7.2 LCA resources

THIS BOOK IS DESIGNED to provide an overview of LCA fundamentals appropriate for building industry professionals looking to understand and interpret LCA results, and is an introduction to the details necessary to conduct an LCA. While basic LCA studies can be useful in aiding design and construction decisions, most building industry professionals will either rely on LCA tools specifically developed for the building industry or identify an individual with LCA expertise to conduct the LCA. Preparing a unique comprehensive LCA requires professional expertise in LCA standards and methods beyond the scope of this book.

Good sources for additional LCA guidance, developing LCA tools and data sources are outlined below with detailed information provided in the references section of the book. Given the fast developing nature of LCA data, standards and tools, readers should use these references as a starting point to identify current best resources.

The Ecology of Building Materials (Berge. B., 2009)
A good overview of the impacts of building materials on environmental ecology including an introduction to environmental profiles and methods to assess environmental impacts as well as a review of the impacts related to typical building materials.

General Guide for Life Cycle Assessment – Detailed guidance & framework and requirements for Life Cycle Impact Assessment models and indicators (ILCD 2010a&b)
Extensive and practical guidance for LCA practitioners on modeling and calculation methodology. A good source for detailed information about specific LCA issues.

The Greenest Building: Quantifying the Environmental Value of Building Re-use (Preservation Green Lab, 2012).

An interesting study using LCA to evaluate the impacts of demolition, renovation and re-use.

Guide to Building Life Cycle Assessment in Practice (AIA, 2010)

A good introduction and resource developed by the American Institute of Architects.

A Life Cycle Approach to Buildings: principles, calculations, design tools (Konig et al., 2010)

An interesting book that provides both an overview of LCA and LCCA as well as advice about how to integrate LCA and LCCA and examples of LCA in practice.

LCA 101: Life Cycle Assessment Principles and Practices (EPA, 2006)

An overview of LCA phases, terms and interpretation prepared by the US EPA.

Operational guidance for Life Cycle Assessment of the Energy-Efficient Buildings Initiative (Wittstock et al., 2012)

A comprehensive web based resource that includes guidance for both product and whole building LCAs.

Proceedings of the 2012 International Symposium on Life Cycle Assessment and Construction, Nantes, France (Ventura & Roche, 2012)

A wide range of research papers available for free online.

2030 Challenge for Products Information Hub (Building Green, 2013)

A website with extensive links and information related to LCA and EPDs in support of Architecture 2030's Product Challenge.

Athena Impact Estimator for Buildings (ASMI, 2013)

This whole building LCA tool is customized for North America and is applicable to new construction, renovation and additions in all building types. The tool is free for download.

Tally®/Environmental Impact Tool (KieranTimberlake, 2013)

This tool, in Beta testing at time of publication, is a plug in to the building information modeling software REVIT to extract material quantities and connect to LCI databases. Developed in collaboration between an architecture firm (KieranTimberlake) and LCA provider (PE International).

Low Carbon Building Method (Fabre, 2010)

This Excel-based tool is designed to enable users to create whole building embodied carbon footprint (single attribute LCA). Using open source carbon footprint data, the tool allows users to override default values for emissions, transportation mode and distance. Built on the PAS 2050 standards.

GaBi (PE INTERNATIONAL, 2013)

GaBi is a comprehensive LCA tool enabling users to make customized LCA models integrating public LCI databases as well as PE's proprietary LCI datasets.

SimaPro (PRé, 2013)

SimaPro is a comprehensive LCA tool enabling users to make customized LCA models using public LCI databases (Including US LCI and Ecoinvent).

BEES: Building for Environmental and Economic Sustainability (NIST, 2013)

Data and software developed by the US Government that includes LCA results for building materials and products. Although not currently being supported and incomplete, it provides data and methods to assess different products.

Bath Inventory of Carbon and Energy (Hammond and Jones, 2011)

A free assembly of embodied carbon and energy data for building materials with references to original data sources.

CEDA EIO Database (CEDA, 2011)

A private EIO-LCA database for the US.

Ecoinvent Version 3 (Ecoinvent, 2013)

The ecoinvent database is a comprehensive LCI database managed by a nonprofit consortium supported by the Swiss Federal Offices.

EIO-LCA (CMU GDI, 2008)

An open source EIO-LCA database and online interface to access an EIO-LCA model.

US Life Cycle Inventory Database (NREL, 2013)

This open source database is designed to provide a repository for LCI data for the United States. Given that publication to the LCI database is voluntary and that limited support is provided to attain and update the data, many materials and processes are not represented in this database.

eTool (eTool, 2013)

This online LCA software developed in Australia has both open access/free and professional versions that permit users to build up a LCI for a building. The software enables users to modify critical LCA variables such as material life and disposal method.

IMPACT (IMPACT Project, 2013)

IMPACT is a tool developed to integrate LCA, Life Cycle Costing and Capital Costing within a 3D modeling platform. It combines a large construction dataset of environmental impacts and costs (developed by BRE & Willmott Dixon), energy use, HVAC modeling, daylighting and water analysis with lifecycle tools. The LCA analysis is compliant with CEN/TC 350 and approved for use with the UK's BREEAM rating system to achieve (LCA and LCC credits). The LCA dataset is valid for Europe (and potentially usable beyond); the life cycle cost data is UK specific. The datasets could be customized to other regions.

Green Footstep (RMI, 2014)

This online tool calculates the carbon footprint of building proposals enabling users to both use default EIO LCA data and customized LCA results from other LCA tools to integrate with site specific landscape and operational energy impacts.

Bibliography

AIA (American Institute of Architects) (2010) *Guide to Building Life Cycle Assessment in Practice.* Available at: http://www.aia.org/aiaucmp/groups/aia/documents/pdf/aiab082942.pdf (accessed October 2013).

Anderson, J. and Thornback, J. (2012) *A Guide to Understanding the Embodied Impacts of Construction Product.* Available at: http://www.constructionproducts-sustainability.org.uk/fileadmin/client/cpa/documents/Sustainability/Embodied_Impacts_brochure_small_V9.pdf (accessed October 2013).

Arup (2012) "Embodied CO_2 of structural frames", *The Structural Engineer*, May, pp. 33–37.

ASMI (2013) *Athena Impact Estimator for Buildings*, Software, Ontario, Canada: Athena Sustainable Materials Institute (ASMI). Available at: www.calculatelca.com. (accessed October 2013).

Athena (2012) *Athena Impact Estimator*, Athena Sustainable Materials Institute. Available at: http://www.athenasmi.org/our-software-data/impact-estimator/ (accessed December 2013).

Atherton, J. (2007) "Declaration by the metals industry on recycling principles", *International Journal of Life Cycle Assessment*, 12(1): 59–60.

Bardi, U. (2009) "Peak oil: the four stages of a new idea", *Energy*, 34: 323–326.

Bare, J. (2011) "TRACI 2.0: the tool for the reduction and assessment of chemical and other environmental impacts", *Journal of Clean Technologies and Environmental Policy*, 13(5) 687–696.

Bare, J.C., Norris, G.A., Pennington, D.W. and McKone, T. (2003) "TRACI: the tool for the reduction and assessment of chemical and other environmental impacts", *Journal of Industrial Ecology*, 6(3–4): 49–78.

Basbagill, J.P. (2013) Integration of life cycle assessment and conceptual building design, PhD dissertation, Stanford University.

Berge, B. (2009) *The Ecology of Building Materials*, 2nd edn, Oxford: Elsevier.

BSI (2011) *PAS 2050:2011 Specification for the Assessment of the Life Cycle Greenhouse Gas Emissions of Goods and Services*, London: British Standards Institution (BSI).

Building Green (2013) *2030 Challenge for Products Information Hub.* Available at: http://www2.buildinggreen.com/topic/2030-challenge (accessed September 2013).

CEDA V.4 (2011) *Comprehensive Environmental Data Archive*, Input–Output LCA database. Available at: www.iersweb.com.

CEN (2011a) *FprEN 15643-2:2001 Sustainability of Construction Works: Assessment of Buildings, Part 2: Framework for the Assessment of Environmental Performance*, Brussels: European Committee for Standardization (CEN).

CEN (2011b) *CEN 15978: Sustainability of Construction Works: Assessment of Environmental Performance of Buildings: Calculation Method*, Brussels: European Committee for Standardization (CEN).

CEN (2012) *CEN 15804: Sustainability of Construction Works: Environmental Product Declarations: Core Rules for the Product Category of Construction Products*, Brussels: European Committee for Standardization (CEN).

Chen, C., Habert, G., Bouzidi, Y., Jullian, A. and Ventura, A. (2010) "LCA allocation procedure used as an incitative method for waste recycling: an application to mineral additions in concrete", *Resources Conservation and Recycling*, 54(12): 1231–1240.

CML (2002) *Abiotic Resource Depletion in LCA: Improving Characterization Factors for Abiotic Resource Depletion as Recommended in the New Dutch LCA Handbook*, Leiden, NL: Institute of Environmental Sciences (CML) at the Universiteit Leiden for the Road and Hydraulic Engineering Institute of the Dutch Ministry of Transport, Public Works and Water Management. Available at: http://www. cml.leiden.edu/research/industrialecology/researchprojects/finished/abiotic-depletion-lcia.html#abstract (accessed October 2013).

CMU GDI (2008) *Economic Input-Output Life Cycle Assessment (EIO-LCA), US 1997 Industry Benchmark Model*, Pittsburgh, PA: Carnegie Mellon University Green Design Institute. Available at: http://www.eiolca.net (accessed August 2013).

Cole, R. (2012) "Transitioning from green to regenerative design", *Building Research and Information*, 40(1): 39–53.

CTL (2003) *Life Cycle Inventory of Slag Cement Manufacturing Process*, Sugar Land, TX: Slag Cement Association, Available at: http://www.slagcement.org/Sustainability/pdf/Slag%20Cement%20LCI%20Letter%20Report%20August%2011%202003. pdf (accessed October 2013).

Cullen, J., Allwood, J. and Bambach, M.D. (2012) "Mapping the global flow of steel: from steelmaking to end-use goods", *Environmental Science & Technology*, 46(24): 13048–13055.

DEQ (2010) *A Life Cycle Assessment Based Approach to Prioritizing Methods of Preventing Waste from Residential Building Construction, Remodeling, and Demolition in the State of Oregon*, Phase 2 Report, Version 1.4, Portland, OR: State of Oregon Department of Environmental Quality (DEQ). Available at: http://www.deq. state.or.us/lq/pubs/docs/sw/ResidentialBldgLCA.pdf (accessed October 2013).

Doney, S., Fabry, V., Feely, R, and Kleypas, J. (2009) "Ocean acidification: the other CO2 problem", *The Annual Review of Marine Science*, 1:169–192.

Ecoinvent (2013) *Ecoinvent Version 3.0 Database*, Digital database, Switzerland: Swiss Center for Life Cycle Inventories. Available at: www.ecoinvent.org/database (accessed October 2013).

Ecolabel Index (2012) *Index of Ecolabels*, Website database, Vancouver, CA. Available at: http://www.ecolabelindex.com/ecolabels/ (accessed October 2013).

EdBGuide Project (2013) *Operational Guidance for Life Cycle Assessment Studies in the*

Energy Efficient Buildings Initiative, Part A: Products and Part B: Buildings, Brussels: European Commission for Research and Innovation Environment. Available at: http://www.eebguide.eu/ (accessed October 2013).

EEA (European Environment Agency) (2012) *Environmental Indicator Report 2012: Ecosystem Resilience and Resource Efficiency in a Green Economy in Europe*, Brussels: European Union Publication. Available at: http://www.eea.europa.eu/publications/environmental-indicator-report-2012 (accessed October 2013).

EPA (1999) *Smog: Who Does it Hurt? What You Need to Know About Ozone and Your Health*, EPA-452/K-99-001, Washington, DC: Environmental Protection Agency. Available at: http://www.epa.gov/airnow/health/smog.pdf (accessed October 2013).

EPA (2006) *LCA 101: Life Cycle Assessment Principles and Practices*, EPA/600/R-06/060, Cincinnati, OH: National Risk Management Research Laboratory. Available at: http://www.epa.gov/nrmrl/std/lca/lca.html (accessed October 2013).

EPA (2012) *TRACI 2.1 Tool for the Reduction and Assessment of Chemical and Other Environmental Impacts 2.1*, Database, Research Triangle Park, NC: Environmental Protection Agency (EPA) Office of Research and Development, National Risk Management Research Laboratory, Sustainable Technology Division.

EPA (2013a) *Acid Rain*, Washington, DC: EPA. Available at: http://www.epa.gov/acid-rain/ (accessed October 2013).

EPA (2013b) *Climate Change: Basic Information*, Washington, DC: EPA. Available at: www.epa.gov/climatechange/basics/ (accessed October 2013).

EPA (2013c) *Waste Identification*, Washington, DC: EPA. Available at: http://www.epa.gov/osw/hazard/wastetypes/wasteid/ (accessed October 2013).

EPA (2013d) *Characteristic Wastes*, Washington, DC: EPA. Available at: http://www.epa.gov/osw/hazard/wastetypes/characteristic.htm (accessed October 2013).

EPA (2013e) *Coal Combustion Residuals: Proposed Rule*, Washington, DC: EPA. Available at: http://www.epa.gov/epawaste/nonhaz/industrial/special/fossil/ccr-rule/index.htm (accessed October 2013).

EPA (2013f) *Particulate Matter*, Washington, DC: EPA. Available at: http://www.epa.gov/pm/ (accessed October 2013).

EPA (2013g) *Inventory of U.S. Greenhouse Gas Emissions and Sinks, 1990–2011*, EPA 430-R-13-001. Available at: http://www.epa.gov/climatechange/ghgemissions/usinventoryreport.html (accessed October 2013).

EPA (2013h) *Nutrient Pollution*, Washington, DC: EPA. Available at: www2.epa.gov/nutrientpollution (accessed October 2013).

Ercin, A.E. and Hoekstra A.Y. (2012) *Carbon and Water Footprints: Concepts, Methodologies and Policy Responses*, Paris: United Nations Educational, Scientific and Cultural Organization. Available at: http://www.waterfootprint.org/Reports/Ercin-Hoekstra-2012-Carbon-and-Water-Footprints.PDF (accessed October 2013).

eTool (2013) *eTool*, Austrailia, eTool. Software. Available at http://etool.net.au/ (accessed February 2014).

EU (European Union) (2008) "Directive 2008/98/EC of the European Parliament and of the Council on Waste and repealing certain directives", *Official Journal of European Union*, L312: 3–30.

EUCE (European Commission for the Environment) (2013) *Consultation on the Review of Hazardous Properties*, EU: Waste Framework Directive. Available at: http://ec.europa.eu/environment/waste/framework/pdf/Technical_proposal_tc.pdf (accessed October 2013).

Fabre, G. (2010) *Low-Carbon Buildings Method*, Online tool. Available at: www.lcb-method.com (accessed October 2013).

Fabry, V., Seibel, B., Feely, R. and Orr, J. (2008) "Impacts of ocean acidification on marine fauna and ecosystem processes", *International Council for the Exploration of the Sea Journal*, 65(3): 414–432. Available at: http://icesjms.oxfordjournals.org/content/65/3/414.full.pdf+html (accessed October 2013).

Fahey, D. and Hegglin, M. (2010) *Twenty Questions and Answers About the Ozone Layer: 2010 Update*, Nairobi, Kenya: United Nations Environment Programme. Available at: http://ozone.unep.org/Assessment_Panels/SAP/Scientific_Assessment_2010/SAP-2010-FAQs-update.pdf (accessed October 2013).

Frischknecht, R. (2010) "LCI modeling approaches applied on recycling of materials in view of environmental sustainability, risk perception and eco-efficiency", *International Journal of Life Cycle Assessment*, 15: 666–671.

Fuller, S. (2010) *Life-Cycle Cost Analysis (LCCA)*, Washington, DC: National Institute of Building Sciences Whole Building Design Guide. Available at: http://www.wbdg.org/resources/lcca.php (accessed October 2013).

Guinee, J.B., Heijungs, R., and Huppes, G. (2011) "Life Cycle Assessment: past, present, and future", *Environmental Science and Technology*, 45: 90–96.

Hammond, G. and Jones, C. (2011) *The Inventory of Carbon and Energy (ICE)*, Bracknell, UK: BSRIA, Available at: http://www.constructionstudies.ie/modules/wt4106-materials-tech-/inventory-of-carbon-and.pdf (accessed October 2013).

Hauschild, M.Z., Goedkoop, M., Guinee, J., Heijungs, R., Huijbregts, M., Jolliet, O., et al. (2012) "Identifying best existing practice for characterization modeling in life cycle impact assessment", *International Journal of Life Cycle Assessment*, 18: 683–697.

Hauschild, M.Z., Huijbregts, M.A.J., Jolliet, O., MacLeod, M., Margni, M.D., van de Meent, D., et al. (2008) "Building a model based on scientific consensus for life cycle impact assessment of chemicals: the search for harmony and parsimony", *Environmental Science and Technology*, 42(13): 7032–7037.

Heijungs. R. and Suh, S., (2001) *The Computational Structure of Life Cycle Assessment (Eco-Efficiency in Industry and Science)*, Dordrecht: Kluwer Academic Publishers.

Hendrickson, C.T., Lave, L.B. and Matthews, H.S. (2006) *Environmental Life Cycle Assessment of Goods and Services*, Washington, DC: Resources for the Future Press.

Hoekstra, A., Chapagain, A., Aldaya, M. and Mekonnen, M. (2011) *The Water Footprint Assessment Manual: Setting the Global Standard*, London: Earthscan. Available at: http://www.waterfootprint.org/?page=files/WaterFootprintAssessmentManual (accessed December 2013).

HPD (2012) *Health Product Declaration Standard 1.0*, Sommerville, MA: Health Product Declaration Collaborative. Available at: http://www.hpdcollaborative.org/ (accessed October 2013).

HQE (2012) *HQE Performance: Premières tendances pour les bâtiments neufs* [First trends for new buildings], Paris: HQE Association. Available at: http://www.

businessimmo.com/system/datas/23262/original/Brochure_HQE_Performance. pdf?1333028660 (accessed October 2013).

Humbert, S., Marshall, J.D., Shaked, S., Spadaroa, J.V., Nishioka, Y., Preiss, P., *et al.* (2011) "Intake fraction for particulate matter: recommendations for life cycle impact assessment", *Environmental Science & Technology*, 45: 4808–4816.

Hunt, R.G. and Franklin, W.E. (1996) "LCA: How it came about", *The International Journal of Life Cycle Assessment*, 1(1): 4–7.

ICC (2012) 2012 *International Green Construction Code*, Washington DC: International Code Council.

ILCD (2010a) *General Guide for Life Cycle Assessment: Detailed Guidance: International Reference Life Cycle Data System (ILCD) Handbook*, Luxembourg: European Commission Joint Research Centre Institute for Environment and Sustainability. Available at: http://lct.jrc.ec.europa.eu/pdf-directory (accessed October 2013).

ILCD (2010b) *Framework and Requirements for Life Cycle Impact Assessment Models and Indicators: International Reference Life Cycle Data System (ILCD) Handbook*, Luxembourg: European Commission Joint Research Centre Institute for Environment and Sustainability. Available at: http://lct.jrc.ec.europa.eu/pdf-directory (accessed October 2013).

ILFI (2012) *Living Building Challenge 2.1*, Seattle, WA: International Living Future Institute. Available at: http://living-future.org/lbc (accessed October 2013).

IMPACT Project (2013) *IMPACT*, United Kingdom: IMPACT. Software. Available at http://www.impactwba.com/ (accessed February 2014).

ISO (1999a) *ISO 14024 Environmental Labels and Declarations: Type I Environmental Labeling: Principles and Procedures*, Geneva: International Standards Organization.

ISO (1999b) *ISO 14021 Environmental Labels and Declarations: Self-Declared Environmental Claims (Type II Environmental Labeling)*, Geneva: International Standards Organization.

ISO (2001) *ISO 14020 Environmental Labels and Declarations: General Principles*, Geneva: International Standards Organization.

ISO (2006a) *ISO 14040 Environmental Management: Life Cycle Assessment: Principles and Framework*, Geneva: International Standards Organization.

ISO (2006b) *ISO 14044 Environmental Management: Life Cycle Assessment: Requirements and Guidelines*, Geneva: International Standards Organization.

ISO (2006c) *ISO 14025 Environmental Labels and Declarations: Type III Environmental Declarations: Principles and Procedures*, Geneva: International Standards Organization.

ISO (2007) *ISO 21930 Sustainability in Building Construction: Environmental Declarations of Building Products*, Geneva: International Standards Organization.

Jelinski, L.W., Graedel, T.E., Laudise, R.A., McCall, D.W. and Patel, C.K.N. (1992) "Industrial ecology: concepts and approaches", *Proceedings of the National Academy of Sciences of the United States of America*, 89: 793–797.

Johnson, L., Lippke, B., Marshall, J. and Comnick, J. (2005) "Life-cycle impacts of forest resource activities in the Pacific Northwest and Southeast United States", *Wood and Fiber Science*, 37: 30–46.

JRC (2011) *Supporting Environmentally Sound Decisions for Construction and Demolition (C&D) Waste Management: A Practical Guide to Life Cycle Thinking (LCT) and Life Cycle Assessment (LCA)*, Brussels: European Commission Joint Research Centre (JRC).

Available at: http://lct.jrc.ec.europa.eu/pdf-directory/D4B-Guide-to-LCTLCA-for-C-D-waste-management-Final-ONLINE.pdf (accessed October 2013).

KieranTimberlake (2013) Tally® Environmental Impact Tool. Available at: http://ktrg.kierantimberlake.com/tally/ (accessed December 2013).

Konig, H., Kohler, N., Kreisig, J. and Lutzkendorf, T. (2010) *A Life Cycle Approach to Buildings*, Munich, Germany: Institut für international Architektur-Dokumentation GmbH & Co.

Krueger, A. and Minzner, R. (1976) "A mid-latitude ozone model for the 1976 U.S. Standard atmosphere", *Journal of Geophysical Research*, 8(24): 4477–4481.

Lamlom, S. and Savidge, R. (2003) "A reassessment of carbon content in wood: variation within and between 41 North American species," *Biomass and Bioenergy*, *(24)*4: 381–388.

Leontief, W. (1986) *Input-Output Economics*, New York: Oxford University Press.

McDonough, W., Braungart, M. and Hoye, S. (2002) *Cradle to Cradle: Remaking the Way We Make Things*, New York: North Point Press.

Meadows, D.H., Meadows, D.L., Randers, J. and Behrens, W.W. (1972) *The Limits to Growth*, New York: Universe Books.

Molina, M. and Rowland, F. (1974) "Stratospheric sink for chlorofluoromethanes: chlorine atom-catalyzed destruction of ozone", *Nature* 249: 810–812.

Moore, E. (2013) "Carbon goggles and other tools for viewing building materials for design", *Proceedings of the Building Technology Educators Society 2013 Conference*, Bristol, RI.

Morison, J., Matthews, R., Miller, G., Perks, M., Randle, T., Vanguelova, E., White, M. and Yamulki, S. (2012). *Understanding the Carbon and Greenhouse Gas Balance of Forests in Britain: Forestry Commission Research Report*, Edinburgh: Forestry Commission.

Newman, P., Oman, L., Douglass, A., Fleming, E., Frith, S., Hurwitz, M., *et al.* (2009) "What would have happened to the ozone layer if chlorofluorocarbons (CFCs) had not been regulated?", *Atmospheric Chemistry and Physics*, 9, 2113–2128. Available at: http://atmos-chem-phys.net/9/2113/2009/acp-9-2113-2009.pdf (accessed October 2013).

NIST (2013) *BEES: Building for Environmental and Economic Sustainability*, Software, Gaithersburg, MD: National Institute of Standards and Technology (NIST). Available at: http://www.nist.gov/el/economics/BEESSoftware.cfm (accessed October 2013).

NREL (2013) *U.S. Life Cycle Inventory Database*, Golden, CO: National Renewable Energy Laboratory (NREL). Available at: https://www.lcacommons.gov/nrel/search (accessed October 2013).

PE INTERNATIONAL (2013) *Gabi 6*, Software, Germany: PE INTERNATIONAL. Available via www.gabi-software.com (accessed October 2013).

Phister, S., Koehler, A. and Hellweg, S., (2009) "Assessing the environmental impacts of freshwater consumption in LCA", *Environmental Science and Technology*, 43: 4098–4104.

PRé (2013) *SimaPro*, Software, The Netherlands: PRé Consultants. Available at: www.pre-sustainability.com (accessed October 2013).

Prescott, C. (2010) "Litter decomposition: what controls it and how can we alter it to sequester more carbon in forest soils?" *Biogeochemistry*, 101:133–149.

Preservation Green Lab (2012) *The Greenest Building: Quantifying the Environmental Value of Building Reuse*, Seattle, WA: Preservation Green Lab. Available at: http://www.preservationnation.org/information-center/sustainable-communities/green-lab/lca/The_Greenest_Building_lowres.pdf (accessed October 2013).

Puettmann, M., Oneil, E., and Johnson, L. (2013) *Cradle to Gate Life Cycle Assessment of Glue-Laminated Timbers Production from the Pacific Northwest*, Seattle, WA: Consortium for Research on Renewable Industrial Materials (CORRIM). Available at: http://www.corrim.org/pubs/reports/2013/phase1_updates/index.asp (accessed October 2013).

RMI (2014) *Green Footstep*, United States: Rocky Mountain Institute. Online Software. Available at http://www.greenfootstep.org/ (accessed February 2014).

Rowland, F. and Molina, M. (1994) "Ozone depletion: 20 years after the alarm", *Chemical and Engineering News*, 723: 8.

TerraChoice (2007) *Certification Criteria Document CCD-020: Gypsum Wallboard, Environmental Choice Program*, Ottawa, CA. Available at: http://www.ecologo.org/common/assets/criterias/CCD-020.pdf (accessed August 2013).

UN (2012) *Managing Water under Uncertainty and Risk: The United Nations World Water Development Report 4, Volume 1*, Paris: United Nation Educational, Scientific and Cultural Organization. Available at: http://unesdoc.unesco.org/images/0021/002156/215644e.pdf (accessed October 2013).

UNECE (United Nations Economic Commission for Europe) (2012) *The 2012 Amendments to the Gothenburg Protocol to Abate Acidification, Eutrophication and Ground-level Ozone*, Geneva: United Nations Economic Commission for Europe, Available at: http://www.unece.org/env/lrtap/multi_h1.html (accessed October 2013).

United States Green Building Council (USGBC) (2013) *LEED Rating Systems*. Available at: http://www.usgbc.org/leed/rating-systems (accessed December 2013).

van Oers, L., de Konigh, A., Guinee, J.B. and Huppes, G. (2002) *Abiotic Resource Depletion in LCA: Improving Characterization Factors for Abiotic Resource Depletion as Recommended in the New Dutch LCA Handbook*, Utrecht, NL: Road and Hydraulic Engineering Institute of the Dutch Ministry of Transport, Public Works and Water Management. Available at: http://media.leidenuniv.nl/legacy/report%20abiotic%20resource%20depletion.pdf (accessed October 2013).

Ventura, A. and de la Roche, C. (2012) *International Symposium on Life Cycle Assessment and Construction: Civil Engineering and Buildings*, Nantes, France: International Society for Industrial Ecology and CSTB (the French Scientific and Technical Centre for Buildings). Available at: http://lca-construction2012.ifsttar.fr/ (accessed October 2013).

Vizcarra, A.T., Lo, K.V. and Lioa, P.H. (2009) "A life-cycle inventory of baby diaper subject to Canadian conditions", *Environmental Toxicology and Chemistry*, 12: 1707–1716.

Wittstock, B., Gantner, J., Lasvaux, S., Saunders, T., Fullana-i-Palmer, P., Mundy, J.A. and Sjöström, C. *et al.* (2012) *Operational Guidance for Life Cycle Assessment of the Energy-Efficient Buildings Initiative, Eebguide Guidance Document Project Report*, Stuttgart: Fraunhoffer Institute for Building Physics. Available at: www.eebguide.eu (accessed October 2013).

WN (World Nuclear) (2013) *Naturally-Occurring Radioactive Materials*, London: World Nuclear Association. Available at: http://www.world-nuclear.org/info/

Safety-and-Security/Radiation-and-Health/Naturally-Occurring-Radioactive-Materials-NORM/#.UfAwEGQ4WI0 (accessed October 2013).

World Resources Institute (WRI) (2005) *Navigating the Numbers: Greenhouse Gas Data and International Climate Policy*, Washington, DC: World Resources Institute. Available at: http://pdf.wri.org/navigating_numbers.pdf (accessed October 2013).

WRI/WBCSD (2011) *GHG Protocol: Product Life Cycle Accounting and Reporting Standard*, Washington, DC: World Resources Institute and World Business Council for Sustainable Development (WBCSD). Available at: http://www.ghg-protocol.org/standards/product-standard (accessed October 2013).

WSA (2011) *Life Cycle Assessment Methodology Report*, Brussels: World Steel Association (WSA). Available at: http://www.worldsteel.org/steel-by-topic/life-cycle-assessment/about-the-lci.html (accessed October 2013).

Yellishetty, M., Mudd, G.M. and Ranjith, P.G. (2011) "The steel industry, abiotic resource depletion and life cycle assessment: a real or perceived issue?", *Journal of Cleaner Production*, (19)1: 78–90.

Index of Standard Definitions

As noted within the text, there are standard definitions for terms used in LCA. The primary LCA standards (ISO 14040, 14044, 14025 and 21930 as well as CEN 15804) related to buildings and building products provide specific definitions for the terms noted below. The standards are copyright protected and thus not available for reprinting here. However, ISO provides online previews of their standards (which currently includes access to the terms and definitions sections) at their online browsing platform accessible at https://www.iso.org/obp/ui/#home.

Term	Standard definition
abiotic resources	
acidification	
allocation	ISO 14040
ancillary input	ISO 21930
average data	CEN 15804
building product	ISO 21930
by-product	
carbon footprint	
category endpoint	ISO 14040
characterization factor	ISO 14040
comparative assertion	ISO 14040
completeness check	ISO 14040
consistency check	ISO 14040
construction element	CEN 15804
consumer	ISO 14025
co-product	ISO 14040 & CEN 15804

Index